Pelican Books
Noise

Rupert Taylor was born in 1946 in Northumberland.
He was trained in acoustics while with a firm of
acoustical engineers, Burgess Products Co. Ltd,
whom he joined at the age of eighteen. After four
years' work with them he took the plunge of
setting up as an independent Noise Control
Consultant, and little more than two years later he
formed the consultancy firm of Rupert Taylor and
Partners Ltd in conjunction with another acoustical
engineer, John Riding. The business proved
very successful and continues to expand.

He served a term on the Council of the British
Acoustical Society, and has been involved in
consulting assignments ranging from cutting down
noise in power stations and designing vast silencers
to the control of noise from clients in a restaurant.

Married, with no children, he admits to being a
somewhat mediocre musician and perhaps a better
artist, but most 'relaxation' is provided by the New
Forest Ponies which his wife and he breed and show.

Rupert Taylor

NOISE

Penguin Books

Penguin Books Ltd, Harmondsworth,
Middlesex, England
Penguin Books Inc., 7110 Ambassador Road,
Baltimore, Maryland 21207, U.S.A.
Penguin Books Australia Ltd, Ringwood,
Victoria, Australia

First published 1970
Copyright © Rupert Taylor, 1970

Made and printed in Great Britain by
Cox & Wyman Ltd,
London, Reading and Fakenham
Set in Monotype Times

To Alison

Contents

List of text figures

List of tables

Acknowledgements

I am very grateful to Rolls-Royce Ltd and C.A.V. Ltd for the help and material I have received from them, to M.F. Russell and to A.E.W. Austen, T. Priede and the Institution of Mechanical Engineers for permission to use material from their papers on diesel engine noise. Many of the books listed under 'Further Reading' have been most valuable. I thank Eileen Love for a vast amount of typing in a very short time and Hugh Rusden for his assistance. Finally I owe a debt of gratitude to Patrick Coombs, whom I thank for my very presence in the field of acoustics.

Foreword

I am not an academic. I have tried to condense the knowledge gained in six years of controlling noise into a book for fellow non-academics. There is not much mathematics; what there is should not be too much for anyone but is anyway not essential to the meaning of the text.

I have tried to say nothing that cannot be justified in simple English, and have tried to justify some things which would normally be done mathematically. This is not a handbook, but a narrative about the art and the state of the art of acoustics. No, it is no longer just an art, it has blossomed into science, and it has been a science for much longer than people think.

It would be vain of me to worry that I had given enough away to be bad for business!

1 Acoustics – the unknown science

The phenomenon of sound is as old as the earth itself. The turmoil that accompanied the formation of this planet created shocks, vibration and sound, no doubt of immense magnitude. When the earth cooled and life began, nature continued to make sounds, of waves beating on rocks, of wind howling through trees and thunder in the sky. Nature can still make as much noise as man; the noise of the eruption of Krakatoa which in 1883 blew up about a cubic mile of rock to rise as dust over fifteen miles in the atmosphere would equal or surpass that of a modern nuclear explosion.

The creatures which evolved to live on the newly cooled planet developed extensions of their brains which enabled them to perceive this phenomenon in the air around them. The sense of hearing gave them an increased chance of survival; sound carries information, and the information it carried for those early earth dwellers was first of all warning of danger. Later came communication: vocal sounds of various types could be used to pass on sensations of danger, could help to keep animals with herding instincts together and aided in the reproductive process by enabling a member of one sex to find a partner of the other through the mating call.

It was not until man appeared on the scene that any creature made full use of the medium at his disposal. Man's grunts and groans achieved greater and greater variety to cope with the increasing amount of information he needed to convey. As a co-operative, hunting animal, he could no longer lounge in the branches of a tree and reach lazily for a dangling fruit like his primate ancestors, and communications became vital. The location of prey, organizing the help of others in the chase and the kill, and again the warning of impending danger all called for more and more meaningful communication and so speech was born out of severe necessity.

Man soon found that sound had other uses: he noticed that empty cooking vessels made sounds when struck, and that when he drew his hunting bow and released the arrow, the string of the bow twanged and hummed, and thus the earliest musical instruments were evolved.

Music and speech were not man's only contribution to the world of sound. From the earliest days he chipped away at flints to form primitive weapons and many a cave man must have clubbed his neighbour because of the intolerable 'clink, clink, clink' of his axe-making. Then he invented the wheel and thus, unknowingly, sowed the seeds of the noise problem of today. The rumble of wheels on the stones beneath caused insomnia to many an ancient and in later years in towns it became common practice to lay straw in the street outside one's house to deaden the clatter of horses' hooves and the rattle of iron tyres on the cobbled streets.

The coming of the iron age brought new noises to man's ear and the beating and hammering of metal to make weapons and utensils reverberated through the villages. At the same time as man was learning to create pleasurable stimuli to his sense of hearing, in other words to create music, he was beginning to pollute his surroundings and blunt his hearing by making more and more loud and unpleasant crashes and bangs, grindings and rumbles. Primitive societies today are predominated by the sounds of nature and characterized by their lack of man-made sound. It was the growth of mechanization that brought with it the upsurge in noise. Thus for centuries man's ears were sullied by no worse noises than those created in the fashioning of wood, stone and metal. Gunpowder brought with it the comparatively new sound of explosions, and also the first real danger of damaging the hearing. The noisiest occasions of those days no doubt were battles, when the crashing of axes, swords and armour and the booming of cannon were to be heard above the anguished cries of men and the sound of bugles and drums.

It was the Industrial Revolution in Britain which heralded the start of the noise age. The new factories, mines and ironworks brought pollution of all kinds – smells, fumes and eyesores, and of course noise. With the invention of the steam engine and the construction of great machines the noise problem gathered momentum. The spread of the railways, the advent of the internal combustion engine and the increasing use of steel brought ever more noise. Now we have the diesel and the jet engine together with advanced and complex machines all adding to the rising cacophony. We have reached the stage where it is the majority of people who are disturbed by noise. The Wilson Report* compares the results of surveys carried out in 1948 and 1961 in which 1,400 people were asked whether they had ever been disturbed in their homes by external noises. In 1948, 23 per cent of those asked said that they were disturbed by noise, and by 1961 this figure had risen to 50 per cent.

The sad fact is that with few exceptions the advance of engineering and technology has caused a universal increase in noise. It is not very surprising, as nobody really knew anything about the subject until the mid nineteenth century, and the art of noise control was little more than just applied common sense. It was not until the Second World War, when great strides in the development of acoustics were made in America and Europe, that it grew to be the exact science of today. Scientists over the ages have toyed with sound; Pythagoras experimented with the vibration of a stretched string, but for hundreds of years it looked as though his efforts represented not only the beginning but also the end of the study of sound. Some eminent men, from Aristotle and Euclid to Ptolemy, later produced theories, but not directly relating to the physical aspect of sound. It was not until 1,500 years later that the foundations of acoustics were laid by Galileo Galilei, who more or less took over where Pythagoras left off 2,000 years before him and started the ball rolling, or the string vibrating, again.

The fathers of the modern science of acoustics were undoubtedly Hermann Helmholtz and Lord Rayleigh, who in the latter half of the last century independently developed many fundamental theories. Helmholtz is remembered particularly for his work on the theory of resonators, and Rayleigh is famous among other things for his publication in 1877 of *The Theory of Sound*, in which he pro-

*Committee on the Problem of Noise, *Noise*, *Final Report*, H.M.S.O., 1963.

pounded several important new concepts. He received the Nobel Prize for Physics in 1904. In latter years, research into noise and its control at first centred around the field of air and surface transport, together with the work which was done on architectural acoustics, but has now mushroomed to embrace every aspect of the problem. Almost every industrial country now has a society devoted to acoustics, and the list of publications on the subject grows ever longer. The science now covers such diverse fields as ultrasonics, with its great practical value in industry, sonar and underwater acoustics, vibration, musical acoustics and audiology. Several universities teach the subject, and many large industrial organizations have their own acoustical research facilities.

Nevertheless, ask the man in the street to say at all precisely what a decibel is, and you will be very lucky to get an intelligent reply. I once arrived on the scene of a noise problem to be greeted by the client, a qualified engineer, who exclaimed defiantly: 'I know all about noise, you measure it in Jezebels!' On another occasion the proprietor of a factory ordered that microphones be installed throughout the works, and be connected to amplifiers and loudspeakers on a patch of waste land outside so that the noise in the works could be sucked up by the microphones and blown out by the loudspeakers! The very word 'acoustics' to many people symbolizes mysterious phenomena and mystical skills.

Every fifth-form physics class is taught the theory of sound; many a boy has played with tuning forks, knows about wavelengths and frequencies, and the young musician is taught about vibrating strings, pipes and harmonics, but it seldom goes further than that. Many designers of the complex machines of today carry the burden of being responsible for some of the greatest annoyance to a large number of people because of noise, and yet their training contains almost no reference to noise or noise prevention. Only now are the lecture theatres of universities and institutions filling up with men who can no longer get by without any knowledge of acoustics.

Noise control as a serious subject for study is a very late starter in the technological race. Noise seems to have been taken for granted or assumed to be inevitable. For years people shrugged their shoulders and thought it was 'one of those things' that nobody could do much about. Thousands of men and women in the cotton mills have their hearing permanently damaged within a few years of going into the trade; boilermakers tell you, even proudly, that after riveting inside a boiler they cannot hear for a week, but few people have thought of such conditions as more than just occupational hazards.

Noise-induced hearing loss, as deafening by noise is officially described, is so widespread among men that it is very difficult indeed to carry out surveys to establish criteria for 'normal hearing'. Few men who have been on active service have escaped without damage to their hearing from gunfire. Many diesel-engined vehicle cabs are subject to excessively high noise levels and most factories have dangerous noise levels in some part of their works. Young people may be sustaining irreparable damage to their ears from the very high level of sound produced by beat groups and other modern musicians. In general, wherever there is industry there is hearing damage; a sound engineer working for a large cinema chain recently noticed that the amplifier settings in cinemas in the North of England were invariably higher than those in the South.

Why is there so much noise? Is it unavoidable or is it ignorance? Noise is a by-product, but whereas a useless by-product in the chemical industry for instance would represent a loss of profit, and engineers would soon find a way of converting the by-product to good use, the waste of energy through noise emission is so small as to be totally negligible. The leakage of energy in the form of noise from a jet engine, say, developing a total power of several hundred thousand kilowatts is as little as a few tens of kilowatts. On top of this, because of the workings of the human ear, to reduce the noise of a jet engine to inaudibility would require that the amount of energy conversion into sound compared with the total energy output of the engine be restricted to one part in ten thousand million! Some noise is obviously unavoidable, and jet engines will never be whisper-quiet, but the great majority is very much avoidable and the reason why it exists is that either the maker of the noise or the designer of the machine that makes it does not know how or cannot afford to silence it. Very, very few noise problems are insoluble, if there is the money to pay for the solution. Economics is as integral a part of noise control as acoustics and this must never be forgotten. It has been said that the gain in efficiency which has been brought about in industry through automation and the general speeding-up of machines has been almost completely offset by the loss of efficiency of the operators because of the increase in noise. This may be an over-statement, but it is certainly true that the reduction of noise levels in a factory can raise the efficiency of the staff. Much of the cost of silencing today is a result of having to reduce the noise as an after-thought. It can cost twice as much to modify a machine or in-stallation when it is complete as to build quietness into the design in the first place. If a factory reaches the stage when it emits so much noise from a new process that it has to shut down because of an injunction, the loss of production can multiply the cost of silencing many times.

This book cannot help to raise the money, but it does set out to dispel some of the ignorance and point the way towards the carrying-out of economical and practical answers to noise problems. Our civilization is suffering severely from pollution from a great many sources. The rivers are being turned into poisonous sewers; the air is tainted with noxious gases; the ground is accumulating toxic levels of poison from herbicides, insecticides and fertilizers and the whole earth is being contaminated with radioactivity from nuclear fall-out. Pollution of one kind or another may well be the ultimate agent by which the reign of mankind on earth is terminated. Now one of the most valuable of our five sen-ses is being blunted by another pollution, noise. Though not deadly, noise ever present, ever irritating, can change men's lives and even their personalities. Probably vastly more serious than the measur-able physiological effects of noise are the immeasurable psycho-logical effects. Prolonged loss or interruption of sleep can lead to disturbance of the mind; irritation and annoyance can drive people to do unnaturally belligerent acts, even to the point of killing the culprit, and noise in general seldom fails to kill efficiency and inspiration.

Man's advanced state of civilization has been possible largely because of his ability to communicate, and one of the two principal methods of communication is by sound. Noise interferes with communication and this has manifold results. Not only is the quality of life lowered and normal activity hampered, but danger signals and warning shouts can be masked, with fatal results. If a machine goes wrong, very often the first sign is the uncharacteristic noise which may give warning of anything from a breakdown to an explosion. If the noise level in a factory is too high, these signals pass unnoticed.

The noise problem has grown to its immense proportions in a very short time. The rapid introduction of urban motorways, the fast-growing preference for diesel engines and small high-speed petrol engines, the advent of jumbo-jets and supersonic aircraft have all contributed. In industry the automation explosion is heard throughout the factories; the faster anything is done, the more potentially noisy it is. The one thing which is not expanding at anything like the same rate is knowledge about noise and how to avoid it. Every new factory should be designed for safe noise levels for each occupant; every new machine should be designed to operate quietly. Domestic machines could all make less din, every new house in a noisy area should be acoustically treated, every motor car should be built to run quietly.

The priority is without doubt to put an end to the confusion and lack of knowledge. It must no longer be possible to blind with scientific talk of decibels, and acoustics must take its place at the forefront so that design in the future is for quietness. Members of the public must know enough not to be fooled into accepting unnecessary noise. Engineers must learn enough about noise to know when it can be kept down, and to understand what the acousticians are talking about. Economy can no longer be the limiting design factor, and already, in such fields as aero-engine design, machines are being built to a noise level as well as to a price and technical specification. This attitude must spread throughout the whole of industry and government if the future is not to bring Hell on Earth.

2   Air, waves and sound

Noise, of course, is only a particular type of sound. It has become common practice to define it as 'unwanted sound' and to a certain extent this is true; what is noise to the ears of one is music to the ears of another. The roar of aircraft passing overhead is intolerable to the unfortunate who lives in the house below, but the noise of an aircraft returning from its maiden flight would gladden the heart of the test pilot's wife. Nevertheless, all noise is sound of some kind or other, and before embarking on skilful methods of reducing it one must first understand completely what sound really is.

It is surprising how many people who remember what they learnt about sound, and were told about vibrations and waves, have carried with them since the impression that those wavy lines in the textbooks were some sort of 'picture' of a sound wave, and that the air is full of invisible wavy lines spreading out from every sound source. This of course is not so, and we can go no further without taking a very close look at the nature of sound itself.

Sound can occur in many media. We are used to experiencing it in air, but you will also find it in water, concrete and virtually any solid, liquid or gas you like to mention; there can be no sound in a vacuum. So the first step is to get to grips with the characteristics of the medium, and as most sounds we hear come to us through the air, let us take air as an example. In case you are having difficulty in thinking of any sound which does not come to you through air, press the back of your wristwatch to your forehead in quiet surroundings and if your hearing is good the ticking you will hear will have travelled through the metal of the watch and the bones of your head, not through air.

Air is elastic. If this sounds like saying air is like rubber and you know very well it is not, forget about the normal meaning of the word and remember that if you compress a body of air, such as the air in a bicycle pump, it will tend to spring back. That is all the word elastic means; if you deform an elastic body it will resist deformation and want to spring back to its original state. The opposite of elastic is 'plastic'. Again, you must forget about washing-up bowls and think of something plastic as a substance which, when deformed, stays that way. Washing-up bowls made of plastic are so-called because they are made out of one of the 'thermoplastics' which behave in a plastic manner when hot, but not when cold. Any medium, to carry sound, must be elastic. This is because no substance which is inelastic is capable of transmitting wave motion, and this is what sound is all about.

If I wanted to attract your attention, I could throw a stone at you. If you were particularly insensitive and the first stone had no effect, I could throw another. If you were determined to take no notice of me, the chances are that I would run out of stones. Now I could use a much more economical method, and prod you with a stick. The beauty of this is that after each prod I get my stick back, and I can go on until I get either what I want or what I deserve. Strangely enough, the difference between these two methods of torment is fundamental: both ways I can bring up equally large bruises by transmitting just as much force, but when I am throwing the stones I am transmitting not only force, but also a great many stones. The remarkable thing about prodding you with my stick is that there is no net transmission of anything except force. I have managed to bruise your arm, and I have still got my stick. Nothing but energy has passed from me to you in the long run. In exactly the same way, when you hear an aircraft in the distance you are receiving nothing more from it than a minute amount of energy. No air particles are flying from the aircraft to you, and no wavy lines are speeding through the air; this is the unique thing about wave motion.

If I may resume prodding you with my stick, you will notice another thing. The stick oscillates back and forth between you and me. It alternates between touching your arm and being about 100mm away from it, and once again this is a fundamental property of wave motion. However, we still have not seen how elasticity is so essential, and how energy can travel over much longer distances than this and at very high speeds.

Any substance, be it solid, liquid or gas, is made up of millions upon millions of tiny molecules in apparently close proximity. In reality, compared with their own actual size, the molecules are not all that close together, but are kept apart by complicated forces which can be likened to springs. The only difference between solid, liquids and gases is that in solids the molecules are stacked in a more or less rigid pattern and the molecular forces are strong. They are by no means stationary, and are in fact continually bouncing about on the 'springs' which attach them to their neighbours. The hotter the solid, the more agitated the molecules become until melting point is reached and the solid becomes a liquid.* In both the solid and the liquid state the molecules on the surface have no partners to attach to on one side, and the spare energy goes towards forming an extra strong attachment towards the other surface molecules, causing the phenomenon of surface tension. When yet more heat is applied, the agitation of the molecules becomes so great that they cease to have more than a fleeting attachment to any partner, surface tension breaks down and the liquid vaporizes. Evaporation can of course occur below boiling point because there is always a steady flow of molecules accidentally breaking loose at the surface of the liquid. The important thing is that although the molecules of a gas have no fixed relationship with their neighbours they are still surrounded by the complex spring-like forces which, when the molecules are cooled and calm down again, will bind them all together back into a liquid and a solid. They may not care who their neighbours are, but they are very touchy about how close they come.

This explains why solids, liquids and gases can all be elastic: if stress is applied, the molecules will be packed closer together or pulled further apart, and their spring-like bonds will be compressed or expanded. As soon as the stress is removed, they are pulled back to their original equilibrium position by the 'springs'. In gases, they are not so much pulled back to their original positions as pushed back to their original distance from their neighbours.

* Some solids sublime, changing directly from the solid to the vapour state. Examples are iodine and dry ice (solid carbon dioxide).

Imagine a long tube filled with air (figure 1) into one end of which a piston is inserted. If the piston is suddenly moved forward in the tube, you might think that the whole column of air would move along simultaneously in front of the piston to make room for it, but this is not what happens. Air molecules, as well as being spaced from their neighbours by spring-like force, are minute quantities of matter and therefore have mass. Each molecule weighs a tiny amount, and because it has mass it has inertia. Remember Newton's First Law, 'Every body continues in its state of rest or uniform motion in a straight line except in so far as

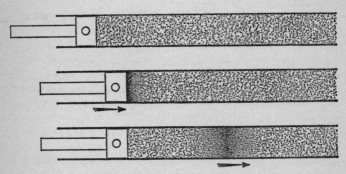

1. The effect of a piston movement on air molecules in a tube

it is compelled by external forces to change that state.' However well oiled the hinges, it will always take an effort to close a heavy oak door, because by virtue of its mass it is reluctant to move. Once it has started to move, it will take almost as much effort to stop it.

On a much smaller scale, the air molecules next to the face of the piston are reluctant to move. The piston jumps forward, and because of the inertia of the first layer of air molecules they them-selves do not move instantaneously; the springs separating them from the face of the piston are therefore compressed. The potential energy given to the springs by the piston eventually sets the molecules moving forward. Because they have mass, and are now moving, the molecules have kinetic energy: the potential energy in the springs has been converted into kinetic energy in the molecules. The process is then repeated when the first layer of molecules tries to push the second layer, which again are reluctant to move because of their inertia and only do so after the springs between the two layers have been compressed. This phenomenon occurs on a much larger scale when a locomotive shunts a train of railway wagons. The inertia of the first wagon, and to a lesser extent friction, causes the springs in the buffers to be compressed, and only when the potential energy in the springs builds up will the wagon start to roll. It will in turn compress the springs between it and the second wagon, and in this way it is quite a time before the last wagon receives any shunt at all.

It is now easy to see why, when the piston suddenly moves a short distance, the air column does not move instantaneously and uniformly – because it takes time for each layer of molecules to get the next layer going. In fact, 334 metres down the tube it would take a whole second before the air would jump forward in the same way as the piston did. If the molecules were made heavier, or the molecular forces weaker, it would take longer, and vice versa. The distance of 334m applies to air at 20°C; at 0°C it would be 332m, to the nearest metre. This is because, as the air cools down, the molecules come closer together, and if it were possible to count the number of layers of molecules in 334m of tube filled with air at 20°C you would find that it would be the same as the number of layers in 332m with air at 0°C. Although in the cold air the molecules are closer together, if the pressure is the same the strength of the springs will be the same, because the molecules are not nearly so agitated. All this applies in exactly the same way if the piston jumps backwards instead of forwards. Instead of the piston pushing on the molecules and compressing the springs, it now pulls and stretches them, before the first layer of molecules will start to move back. They in turn, with their kinetic energy in the reverse direction, pull on the springs between them and the second layer, and so on. Eventually all the air molecules will be back in their original places, the piston having jumped forward and then back again, and they will have transmitted the 'prod', as the stick did, without having made any permanent movement themselves.

The air column is not really all that different from the stick. If you had a method of measuring it you would find that it took a couple of milliseconds for the push on my end of the stick to become a prod on the receiving end. With the air column it takes longer, but the push exerted by the piston at one end certainly becomes a prod at the other end as you would discover if you fixed a delicate membrane across it. Now fit the piston to a crankshaft and start it oscillating to and fro; the delicate membrane the other end will do the same, although out of step or phase because of the time it takes for each movement of the piston to be transmitted from one layer of molecules to the next. Another way of describing what is happening would be to say that the membrane was being oscillated by the *sound* of the piston! However, in reality it would not be possible to make a small enough crankshaft because the actual displacement of air molecules in sound is normally minute. When you hear someone speaking in a normal voice, for instance, the air molecules at your ear are only moving a distance the equivalent of about the diameter of a hydrogen molecule from their rest position.

Unless the tube is very short or the speed of the piston very slow, the effect of each movement of the piston will not have reached the membrane at the other end before the piston has oscillated at least once, if not several times, more. We know that if the tube is 334m long it will take one second for the shunt or sound wave which the piston starts off to be transmitted to the other end. If the piston oscillates at a rate of 100 cycles per second it will alternately push the air 100 times a second and pull it 100 times a second. The time lapse between successive pushes will be 1/100th of a second, during which time the effect of the first push will have travelled 3·34m down the tube. If, therefore, it were possible to 'freeze' the air in the tube and examine it at any instant, we would expect to find a series of compressions at intervals of 3·34m all the way along the tube, and rarefactions (the opposite of compressions) equally spaced between them, as is shown in figure 2.

If instead of 'freezing' the air we could look at one individual molecule, we would see it oscillating to and fro just like the piston. If the molecule we chose to watch were exactly 3·34, 6·68, 10·02 or any multiple of 3·34m away from the rest position of the piston it would be oscillating exactly in phase with the piston. If it were midway between any of these points it would always be doing

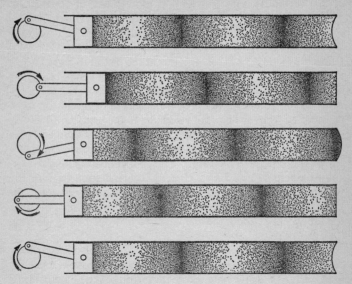

2. The effect of repeated piston movements on air molecules in a tube

exactly the opposite and would be said to be 180° or 3·14 (π) radians out of phase. If you regard the motion of the piston when it has completed a push and a pull as having gone a full circle, you can see that it makes sense to measure any intermediate point as an angle, known as the phase angle, expressed either in degrees or radians. (360° is equal to 2π or 6·28 radians.)

We can now state not only that the membrane at the end of the tube is being vibrated by the sound of the piston, but also that the sound has a frequency of 100 Hertz (abbreviated Hz, meaning 'cycles per second') and a wavelength of 3·34m. The wavy lines referred to earlier now come into their own. They are simply graphs showing the pressure of the air, above or below atmospheric pressure, either at successive points along the tube at any one instant, or at successive instants at any one position. They thus show up the periodic compressions and rarefactions which are sound waves.

Sound in a tube is rather a special case, but the principles are fundamental. Sound in the open air, of course, is not conveniently channelled along a straight line. Instead of a piston and a tube, consider an open space and a small round balloon connected to a pump as shown in figure 3. If air is pumped into and out of the

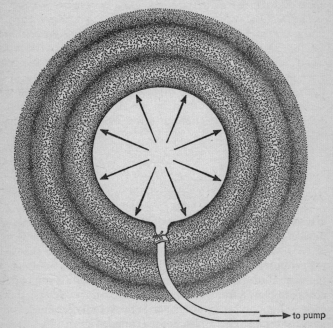

3. The effect on air molecules of a pulsating balloon

balloon, it will of course expand and contract. When it expands, the effect on the air molecules all round it will be similar to the effect on the air molecules in the tube when the piston moves forward. The 'springs' separating the molecules from the balloon are compressed, thus moving the molecules away, which in turn compress the springs between them and the next spherical layer, and so on. When the balloon contracts, the process, as in the tube, is repeated in reverse. The only difference between the two cases in principle is that in the tube the successive layers of molecules which are compressed together or pulled apart are flat and so the compression waves which travel along the tube are 'plane' waves, and with the balloon the layers are spherical. This is, however, an important difference, as we shall see when we come to consider the effect of distance later in the book.

The frequency of a sound has been shown to be directly related to the rate at which the source oscillates. The piston oscillated 100 times a second, causing sound waves at 100 Hz. The wavelength was as much dependent on the rate at which the compression wave travelled down the tube (the speed of sound) as upon the frequency of the source. If the piston speeds up, the wavelength decreases in direct proportion; at 200 Hz it will be 1·67m and at 400 Hz, 0·835m. This relationship can be written out in mathematical terms as follows:

$$\lambda = \frac{c}{f}$$

where λ = wavelength, c = speed of sound and f = frequency.

The frequency of a sound is generally of more interest than the wavelength. It is the frequency which principally determines the pitch of the sound to the listener, as the next chapter will show, and if the wavelength alters at all, for instance because of a change in temperature, the pitch will not alter. A 1,000 Hz note would sound the same at the North Pole as in the Sahara desert. If you had the energy to blow a trumpet in either place, there would be a difference, but only because the change in the velocity of sound would alter the resonant frequencies of the trumpet.

The one thing which will alter the pitch of a fixed-frequency source is a change in the relative velocity of the source and the listener. Most people will have noticed the abrupt drop in the pitch of the engine note as a motor vehicle passes at speed. Both while approaching and receding, the basic engine note in fact remains at the same frequency. Imagine that you are standing close to a conveyor belt which is carrying boxes and is travelling at a speed of one metre per second. A man who is feeding the belt is putting boxes on it at a rate of one per second, so that they will be passing you at a rate of one per second. If, however, he starts to walk towards you at, say, half a metre per second, then the speed of the belt relative to him will go down to half a metre per second. If he continues putting boxes on the belt at a rate of one per second while walking along, he will find that the boxes are spaced only half a metre apart. This means that at your end of the belt the boxes are passing you at twice the rate that he is putting them on. If he were to walk away from you at half a metre per second the reverse situation would occur.

In the same way, if a sound source emits sound waves at a rate of 100 per second, which travel at a rate of 334m/sec, you will hear a 100 Hz note. If the source travels towards you at, say, 33·4m/sec, the wave length of the sound will go down 10 per cent. You will receive 10 per cent more sound waves per second and thus hear a 110 Hz note, and 10 per cent less if it recedes at the same rate with the result that you hear a 90 Hz note. This is called the Doppler effect, and can be most useful in such fields as astronomy, when the Doppler shift in an electromagnetic spectrum can be measured to gauge the velocity of a star.

We can now go no further without finding out more about a very important aspect of a sound wave: its waveform. Let us call in the wavy line, the graph of the sound pressure at any instant at successive points along the path of the wave, or at successive moments at a particular point. Take a sound of a single frequency, say, 1,000 Hz; what do we draw? We know the wavelength if we divide the speed of sound by the frequency, and we know that because it is a single-frequency sound it will consist of a regular alternation between compression and rarefaction. What shape is the graph between these points?

We are looking for the simplest form of repetitive motion. The simplest one that comes to mind is rotation, but it does not solve our problem: movement in a circle will not apply to movement of particles to and fro on a straight line – or will it? If you rotate a bob on the end of a string and look at it from the side, it will not appear to be rotating but simply oscillating up and down. When viewed in this way, the displacement of the bob from the centre varies as the sine of the angle between the string and the horizontal. Motion like this is therefore described as 'sinusoidal', meaning that it varies as the trigonometrical function, the sine.

Sinusoidal oscillation is more commonly described as simple harmonic motion (S.H.M.) and a sound wave propagated by an object oscillating in S.H.M. is the purest type of sound possible, and the only sort of wave which consists of only one frequency. Now we can go back to our wavy line, and it should not be difficult to see that its shape is sinusoidal. There are good, if complicated, mathematical reasons why the most fundamental mode of oscillation or vibration is sinusoidal, and the result is that any elastic body in free oscillation will do so in a manner that will produce sine waves. However, it is very seldom that anything behaves so well that it produces just one, single, unadulterated sine wave, and usually there is a whole series of other lower-amplitude waves added in as well.

Therefore pure tones are rare in practice – nearly all the sounds we hear are very much more complicated, and can consist of a considerable number of tones all lumped together. What effect does this have on the graph? For once, we obtain our answer by simple arithmetic, but to keep up the complexity of the subject it is called 'superposition'. When one is drawing a graph of a sound wave which consists of two or more simple tones or sine waves, one merely adds or substracts the values of each wave at every point and plots the result. You can combine as many waves as you like in this way and end up with some interesting-looking waveforms. The resulting shape will not only depend on the frequencies of the component sine waves, but also on their amplitudes and phase relationships.

Although at first it seems incredible, it is in fact true to say that any waveform, provided it is continuous and repeats itself periodically, can be shown to be made up of a series of sine waves of suitable phases and amplitudes. The man who first demonstrated this was J. B. J. Fourier, a French scientist, and although he was working on the theory of the propagation of heat at the time, his *Théorie Analytique de la Chaleur* published in 1822 became remarkably significant in much wider fields because of the important theorem it contained. Fourier's theorem in full is as follows: 'Every finite and continuous periodic motion can be analysed as a simple series of sine waves of suitable phases and amplitudes.'

We shall find this theorem most useful later in the book, and it is important to understand exactly what it means. The simplest example is that of the square wave, although this sort of waveform occurs in electronics rather than acoustics. Looking at a square wave (figure 4) you would think that, as it looks so totally different from a sine wave, it could not possibly be accounted for as a combination of a series of sine waves, but it can, even if the series is infinite. If, as in figure 4, one starts to build up the components, starting with a sine wave of the same frequency as the square wave and slightly larger amplitude, and adding another of three times the frequency and one third of the amplitude, the square shape begins to emerge. Next comes a sine wave five times the frequency and one fifth of the amplitude of the first sine wave, and by the time the series has got to a wave fifteen times the frequency and one fifteenth of the amplitude of the first wave, the result is beginning to look remarkably like the square wave we are after. It is interesting to notice at this stage that if the waveform to be analysed has sharp curves in it or is jagged, such as the square wave at the corners, the high-frequency components of the Fourier series have to have a much greater amplitude than they do if the waveform is smooth.

In just the same way as it is possible to build up a square wave from a series of sine waves, so is it possible to do the reverse and analyse any other finite, continuous and periodic wave, however unlikely it looks. The full significance of this will appear later.

Air is not the only carrier of sound waves, of course, and virtually any gas, solid or liquid will behave in the same way and transmit sound waves. Although the principles are the same, the quantities such as elasticity and density are vastly different. In most solids, for instance, the speed of sound is at least three times as high as it is in air, because it is proportional to the square root of the ratio of the elasticity, measured in terms of Young's Modulus (E), and the density. If the elasticity is high compared to the density, the velocity of sound will be high; in the case of aluminium the velocity is 5,200m/sec at 15°C or over fifteen times as great as the speed of sound in air. For steel at the same temperature, the velocity is 5,050m/sec, and if you put your ear to the end of a long tube which someone strikes at the other end, you will first hear a bang in that ear due to the wave which travelled in the metal of the tube, and

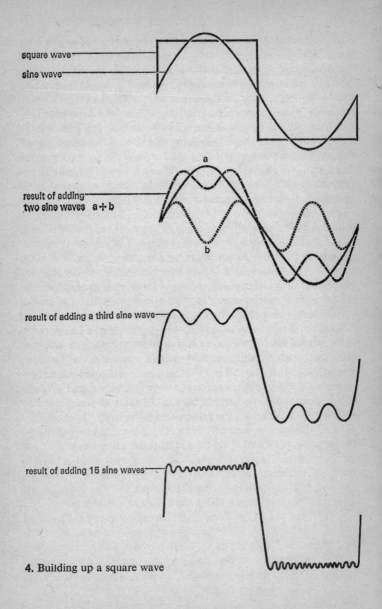

square wave

sine wave

result of adding
two sine waves a + b

a

b

result of adding a third sine wave

result of adding 15 sine waves

4. Building up a square wave

then another bang in the other ear due to the airborne wave which took fifteen times as long to reach you.

Early acousticians used this method to calculate the speed of sound in solids, having worked out the speed of sound in air from timing the lapse between seeing a distant flash and hearing the explosion. They then timed the lapse between the two sounds when a very long bar or tube was struck at the far end. The Frenchman Biot first did this in 1808, using an iron pipe one kilometre long! It was necessary to clamp a bell to the far end in order to make the airborne waves audible over this distance. Two gentlemen named Colladon and Sturm did a similar, less comfortable experiment to measure the speed of sound in water. They submerged a large bell and a charge of gunpowder in Lake Geneva, simultaneously sounding one and igniting the other. They then measured the interval between hearing the bell and seeing the flash at a distance, again under water. In all these experiments the timing device was a stop-watch, and the results were not very accurate. As a matter of interest the speed of sound in pure water has been determined by more sophisticated means to be 1,440m/sec at 15°C.

In later chapters we shall see that solid-borne sound is every bit as important as airborne sound in many noise problems: in buildings, particularly, many sounds make part of their journey through solids. The frequency of a sound in a solid is always the same as it is for the same sound in air, but because of the higher velocity the wavelength is always a great deal longer. One is normally interested in the wavelength in solids only when calculating the resonant frequencies of structures.

3 Hisses, whistles, hums and rumbles

If you feel you are now an expert on pistons in tubes and pulsating balloons, I hope I will not disappoint you too much by saying that you will probably never come across either of them except perhaps in a laboratory, and sound sources are really very much more complicated than that. Although in a book about noise we should perhaps plunge forthwith into the realms of pneumatic drills and supersonic aircraft, the next really logical step to take is to get to grips with some straightforward 'machines' which are not confined to laboratories and which embody most of the noise-making mechanisms of the troublemakers around us: musical instruments. If we are thoroughly familiar with the best way of making sound, we will at least know how not to make machines.

Musical instruments, and many other sound sources, can be put into groups: those that produce sound for reasons of aerodynamics, those whose sound is caused by sustained mechanical vibration, and those that involve impacts or percussion. They make a good example because although they cannot politely be called 'noise sources' (with a few exceptions) it is only because they have been carefully constructed to eliminate all but the required melodious sounds, and are generally played in some sort of harmony with one another (with a few more exceptions). Each one has a less discrimating counterpart in the array of noise-making machines.

Examples from the first group, sources which produce their sound aerodynamically, are certain wind instruments, in particular the flute, recorder and diapason organ pipes. Nothing can produce any sound at all without an energy supply. We have seen that sound is merely a means by which energy, constantly alternating at great speed between potential and kinetic energy, is passed through the air or other medium as pressure waves. With the piston in the tube the energy was supplied by the rotating crankshaft; in the case of the balloon, by the pump. With wind instruments the energy supply comes from the player who supplies a modulated flow of air under pressure from his lungs.

A diagram of a recorder is shown in figure 5. The air under pressure from the lungs enters the mouthpiece through a narrow slit. The air emerges a short distance along as a jet, so causing eddies or vortices to form on alternate sides of the jet. These are created because on both sides of a fast-moving stream of air there is a reduction in pressure. It is in fact because of this that you can lay a penny on a flat surface and flip it over by blowing across the top. The reduction in pressure causes the jet to be sucked out at the sides, and as the air is emerging from the mouthpiece at high velocity and is turbulent, the eddies are formed. The eddies or

5. A diagram of a recorder

vortices next meet the wedge-shaped lip of the recorder and pass either side of it. The lip in fact influences the rate at which the vortices are formed – the shorter the distance between the slit and the lip, the higher the rate of formation of vortices. In the same way, the harder the player blows, the faster the jet of air, and the higher the rate of vortex formation.

The formation of the vortices and their passage on alternate sides of the wedge-shaped lip causes pressure fluctuations in the immediately surrounding air. If you saw off a recorder just past the lip, the sound that would be produced would be a rather unmusical mixture of a hiss and a whistle whose pitch was related to the strength of the blow in the mouthpiece. Although the pressure fluctuations are not nearly as simple, the principle is exactly the same as that of the vibrating balloon. By expanding and contracting, our theoretical balloon was pressing against and pulling back the spherical layer of air around its surface. This caused pressure fluctuations which were passed on layer by layer at the speed of sound in all directions. The only difference between the balloon and the sawn-off recorder mouthpiece is that the latter produces compression and rarefactions, not by the vibration of a surface, but by the vibration of air, and the waves are not neatly spherical.

Why have I sawn off the mouthpiece of my recorder? The answer is that the rest of the instrument has nothing to do with the generation of sound, merely with the modification of the sound which is produced in the mouthpiece. Ignoring for the moment the holes in the side of the body of the recorder, one can see that it is not unlike the tube which contained the piston. There was one important aspect of the tube and the piston which was not mentioned in the last chapter. That is that the compression waves which set off down the tube were assumed to disappear after that. Well, they would not. When a sound wave reaches the end of an open tube it suddenly meets an infinitely large mass of unrestrained air in the world outside, and the effect is rather like that of rolling a small marble against a large one, when the small marble imparts only some of its momentum to the large one, and bounces back. The sound wave in the tube is largely reflected from the open end of the tube. Reflection is one of the cornerstones of acoustics and will be looked at in much more detail later. Having been reflected, the wave travels back down the tube again. In the recorder 'tube' it will soon meet the mouthpiece, and will be reflected yet again, each time losing a bit of its intensity, but the greater part being captive in the instrument.

The pressure at successive moments in the whole tube would look like the graphs in figure 6, which shows the progress of the compression waves down the tube. If we now consider merely the first reflected wave, it will look like figure 6(b), being the reverse of 6(a). Now of course both these waves exist in the tube together, on top of each other, and so to get the overall picture they must be added together by the process called superposition. An amazing thing then occurs. No longer do any of the waves travel in either direction; we are left with one standing wave. No longer does a sound pressure at one place occur at the next place to the right or the left of it at the next instant. Instead, at each point the sound pressure increases periodically, but the amplitude is different in different positions. At point 'a' for example, it is always nil, and at point 'b' it is always at its maximum. With a normal wave the

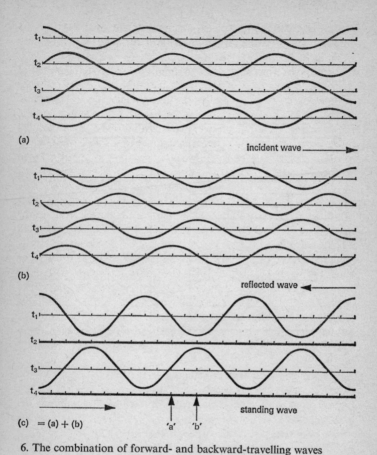

(a)

incident wave ———————➤

(b)

reflected wave ◄———————

standing wave

(c) = (a) + (b)   'a'   'b'

6. The combination of forward- and backward-travelling waves

amplitude periodically reaches the same maximum at all points along the tube, though not all at the same time. As well as being reflected the first and second time, of course the sound wave goes on shuttling backwards and forwards until it is finally dissipated, and when all these reflections are added in, the standing wave pattern is enhanced.

Take another look at the graph of the standing wave; there is another remarkable thing about it. Points of maximum amplitude occur exactly at the ends. This was no accident; it is because the length of the tube is an exact multiple of the length of the sound wave in it. If the wavelength alters or the tube alters in length, figure 7 shows what happens. The standing wave no longer stands,

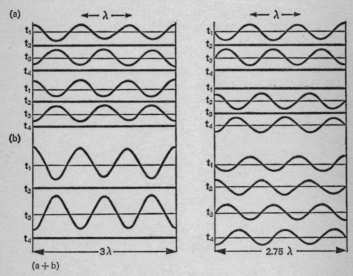

7. The combination of two standing waves

but becomes a travelling wave again of nothing like its previous amplitude. We can now see that the tube is only resonant to sound whose frequency gives it a wavelength that fits exactly into the tube, as in figure 6(c). When the tube is in resonance, the sound which is captive in the tube combines constructively, with the result that the sound pressure is amplified.

The holes in the side of the recorder can now be uncovered. The standing waves were caused because of reflection of the wave from each end of the tube, and at the bottom end of the recorder it is the abruptness with which the wave meets the air mass outside that causes the reflection. It is in fact possible to create this reflection farther back up the tube by uncovering a hole in the side so that the wave meets the air mass outside sooner. This of course effectively shortens the length of the tube, and thus raises its resonant frequency.

Now let us fit the mouthpiece back on to the body of the recorder, and blow. Anyone who has played a wind instrument will know that putting it in the mouth and blowing does not necessarily produce anything like the sound the craftsman who made it intended. Let us cover all the holes, and assume that this should result in the recorder (a treble recorder) sounding the note F above middle C. Turning again to our friends the vortices, these of course will, if the lip of the recorder is fixed in the right position, be formed at a rate related to the strength which is put into the air jet. The body of the recorder is now a tube which is resonant at a frequency of 349 Hz. If from experience the player can blow at the right rate to produce somewhere in the region of 349 vortices per second, the pressure fluctuations thus caused will start the air in the tube oscillating almost in the same way as a piston would. Being resonant, the tube will amplify any wave which travels along it at 349 Hz, having a wavelength of just under one metre.

The lowest frequency at which a closed tube will be resonant is that of a sound whose wavelength is twice the length of the tube. Our instrument in this particular case is therefore just under half a metre long. The resonance in the tube will have a feedback effect and cause the vortices to be formed more exactly at the required rate providing the player is not blowing too hard or too softly. If holes in the recorder body are uncovered the resonant frequency rises, but other than that the process is the same, and the player subconsciously modifies his rate of blowing to create faster vortex formation, helped by the feedback effect. If he starts to blow too hard, first of all the recorder's note will rise in frequency slightly. This is because if the rate of vortex formation is close to, but higher than, the resonant frequency of the body of the instrument, the resulting resonance will be a compromise, midway between the two. If he blows still harder the recorder will scream in protest (what happens is rather more complicated and will be returned to later in the chapter).

Let us now take a look at the members of the other groups and see how they compare. The second group includes the violin family, and again the violin embodies many potential noise-making principles.

It may seem odd, but the easier thing is to make a direct comparison between the violin and the recorder, although they are in different groups. The strings of the violin can be likened to the body of the recorder, the resonant tube. In the resonant tube the initial disturbance was created at one end, but it need not have been, and could quite easily have been transferred to a point along the length of the tube. A stretched string is resonant. You have only to stretch a rubber band and twang it to discover this. It will not, on the other hand, make much of a sound on its own because it is too fine to be able to create large pressure fluctuations in the surrounding air. Figure 8 shows that when a string moves to one side the air tends to slip round it rather than be compressed. This is, however, incidental and does not detract from the fact that the string is resonant.

You will remember that the basic process in the physics of a sound wave consists of a continuous conversion of energy from kinetic energy to potential energy. This also happens in the vibrating string: as you pluck the string and in so doing draw it to one side, you give it potential energy by stretching it. When you let go, the tension in the string tends to restore it to its equilibrium

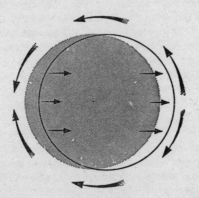

8. The airflow around an oscillating string

position, but the string then has kinetic energy or momentum and this carries it through the rest position and on the other side, back to a position of potential energy, and so on. The string will go on oscillating to and fro until all the energy is dissipated.

There is, however, another more scientific way of looking at the vibration of a string, and this is where the comparison with the tube comes in. If you imagine a very long string under light tension and pluck it at one end, the displacement caused by the pluck will travel along the string rather like a sound wave in a long tube. Similarly, when the pluck reaches the end of the string it will be reflected and travel back again. If the string is forcibly oscillated instead of plucked, the reflected wave will combine with the initial wave and the string will look like a moving graph of a standing wave in a tube. When all the subsequent reflections from the ends of the string are taken into account, one can see how the string will resonate like the tube, but instead of there being points of high pressure and low pressure, the antinodes and nodes, as they are called, are represented by points of high displacement and nil displacement. The resonant frequency is again proportional to the length of the string.

In the same way as you could vary the resonant frequency of the tube by changing the density or elasticity of the gas in it and altering the velocity and therefore the wavelength of the sound, you can do a similar thing with the string. The analogy with the string is that by raising the mass or lowering the tension on the string, or by using a more elastic material for it, the resonant frequency is lowered for the same length of string, and vice versa. This was not realized until the seventeenth century, and so the modern pianoforte's development was severely held back. A piano with strings of equal tension and mass would have to be over twelve metres long to retain the range of seven and a half octaves possessed by modern instruments.

The counterpart of the vortex formation in the violin is of course the vibration created in the strings by drawing the rosin-treated bow across them. There is also the same feedback effect which, once the note is sounded, adjusts the vibration of the string under the bow to coincide with the frequency of the note. Many will know to their acute discomfort that there is more to playing a violin than holding it under the chin and bowing it. The reason for the agonized sounds that the novice fiddler elicits from his instrument is now easier to understand: unless the initial disturbance in the strings created by the bow is made to coincide with the resonant frequency of the string fairly well, and this depends on the pressure and speed of the bow, the resultant sound will have much in common with the squeal which the recorder makes when blown too hard.

The strings, as has been said, cannot make much sound themselves. They do, however, vibrate fairly energetically. This vibration has the ingredients necessary to produce sound – energy and oscillation – but poor means of getting these ingredients into the air. This problem is solved by the body of the violin; the vibration travels via the bridge into the wood of the body which then acts like the theoretical balloon, except that it is working on the air both inside and outside. The sound waves produced inside are partially captive and combine with one another. However, the instrument-maker should have designed the body so that its resonant frequency is lower than the lowest note of any of the strings, otherwise one note would get preferential treatment over all the others. Occasionally he does not succeed, and the result is what is known as the wolf note. The amplification in the violin body is achieved by virtue of the much greater area which is going to work on the air compared with the area of a string on its own. Resonances are the bugbears of bad violins.

Now to the third group: sources dependent on impacts and percussion. The mind jumps to thoughts of cymbals and drums, so let us take cymbals as an example; drums will come into their own in chapter 8. A cymbal has a lot in common with a vibrating string. In theory a vibrating string is a two-dimensional system; the only dimensions we are really interested in are the length and the distance it is displaced from its rest position. A cymbal is in many ways no more than a three-dimensional string. You can set a string vibrating by plucking or even striking it, as well as by bowing it. The same applies to a cymbal, and although you do not often see people bowing cymbals, you may have seen entertainers bowing the edges of sheets of steel and obtaining para-musical sounds.

When you strike a cymbal, bending waves spread out in the metal which are reflected from the edges in very much the same way as happens in the string. The cymbal is not, of course, under tension, but is subject to the elasticity of its material instead. It does have a great advantage over the string in that it is much more capable of turning its vibrating energy into sound, because it is a large flat surface and only the air at the edges is able to slip round and escape compression or rarefaction.

There is still one aspect of sound generation of very great importance which has not yet been touched on. We have seen what sound is, how it is created and how it can be amplified, but although recorders, violins and cymbals have all been described and compared there has been no explanation of the fact that they do not sound remotely alike. They have also been described as excellent examples of sound-generating mechanisms which are to be found in noisy machines everywhere, and yet however many violins were played however badly and loud they would never sound like a pneumatic drill!

To explain all this we must first of all return to the vibrating string, attached to a sounding board to make it easily audible. If while it is vibrating it is touched lightly in the centre, the note it is producing will be heard to rise a complete octave. At first sight the reason for this may seem simple: you are halving the length of the string and thus halving the wavelength and doubling the frequency of the note. The correct explanation, though, is that your finger is preventing the string from vibrating at its resonant or fundamental frequency, and you have discovered that it was all the time vibrating at a whole lot of other frequencies at the same time, known as harmonics. In fact, all the examples so far described have been rather artificial, because there are really very few sounds indeed which are made up of only one single frequency.

We saw earlier that tubes and strings resonated at frequencies governed by their length, because the nodes and antinodes occurred exactly at the ends or centres of the tubes or strings. This is true, but resonances can also occur just as easily at one, two or any integral multiple of the fundamental frequency, because although there are then many more antinodes and nodes, the pattern of them fits just as neatly into the tube or string, as figure 9 shows.

The result is that every musical note, with rare exceptions,

9. Arrangements of nodes and antinodes in resonant pipes (the series is infinite)

consists not only of its fundamental frequency, but also of several, often a multitude, of overtones known as harmonics. Every musical instrument produces notes having their own distinct timbre or tone-colour because of the difference in the number or relative amplitude of the harmonics. Sometimes these are not only the results of multiple resonances of the air column or string, but also of the body of the instrument itself. Wood does not vibrate very easily, because most of the energy is soon dissipated in the material as a result of internal damping. Wooden bells are thus not much good, and woodwind instruments sound fairly pure; there are not any significant contributions from the body of the instrument, and also the rough internal surface, compared with that of brass, for instance, helps to damp out the higher harmonics which give brightness and even harshness to a sound.

Brass, on the other hand, is the opposite of wood, and bronze is a common metal used in bell founding. It also presents a hard, smooth surface inside the instrument so that most of the harmonics remain in the resultant note. The effect of a mute on a trumpet is to damp out many of the harmonics and soften the timbre of the note. Violins can also be muted, by fitting a clamp over the bridge to introduce friction on the strings, rendering the tone soft and silvery. Damping and absorption are looked at in detail later in the book.

The harmonics in the cymbal are very much more complicated than in strings and tubes because of the extra dimension. There are many more permutations of nodes and antinodes, and the geometry of the cymbal can result in there being many resonances of frequency close to one another, which gives this instrument its very colourful, rather non-musical sound.

Now, of course, the explanation of the high-pitched squeals and squeaks which some instruments give out when blown or bowed too hard is obvious. The frequency of the initial disturbance becomes too high to excite the fundamental frequency of the resonator, and instead it excites one of the harmonics. This effect can be turned to advantage, and in fact without it instruments of the horn family would not be possible. In their case, the player can sound the fundamental frequency, sometimes called the first harmonic, or he can tighten his lips, whose vibration is the counterpart of the vortex formation in the recorder, and sound the second harmonic, or the third and many more. Because of the wide spacing in pitch between the lower harmonics, as figure 10 shows, tunes on a bugle, for instance, are restricted to the 'come to the cookhouse door, boys' type, full of intervals of fifths and fourths. A trumpet overcomes this problem because it is really three bugles in one, sharing the same mouthpiece and flare, but with three tubes in between of different length, which can be opened or closed by

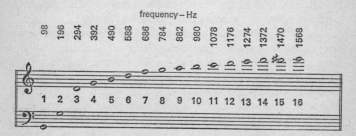

10. Harmonics in musical notation (just temperament). The frequency is 2, 3, 4 etc. times the fundamental

the use of valves. The number of harmonics available is thus greatly increased.

Some confusion exists on this subject of harmonics, because in music, as opposed to acoustics, the second harmonic is sometimes called the first harmonic, the third called the second, and so on. However, throughout this book, the first harmonic will always be the fundamental frequency, and the series will start from there.

So far, we have talked about nothing but pleasant sounds, about recorders, violins and cymbals, but few noises are either melodious or harmonious. Why? What is harmony, or discord? What is it that can make music pleasant and noise ugly? We have seen that a musical note contains most of its energy at the fundamental frequency, with less energy in a range of harmonics. If two identical notes are sounded together, although the vibration may be out of phase, the combination of sound waves is completely regular, both sounds causing pressure fluctuations at exactly the same rate. If the frequency of one note is raised, say from 300 Hz to 350 Hz, the resultant combination will be one note vibrating 300 times, another 350 times a second. If the two are combined, as in figure 11, you can see that the two sounds combine constructively only 50 times a second, or at a frequency equal to the numerical difference between the frequencies of the two notes. The result, although you may not consciously hear it, is that there is a new third note formed, known as a difference tone, having a frequency of 50 Hz. It is the presence of this third, throbbing note which gives this particular combination of notes its characteristically dissonant sound. In fact, when the harmonics of the two notes and their difference tones are added in, the scene becomes quite crowded.

Now if the second note is raised to twice the frequency of the first, so that they are an octave apart, let us see what happens. The frequency of the two notes will be 300 Hz and 600 Hz, with a difference tone at 300 Hz. The difference tone will therefore be at the same frequency as the first note, and will blend in with it as well as the two notes did when they were both at the same frequency. This is why an octave is such a harmonious interval, in fact the most perfect harmony possible. If you work out all the harmonics and their difference tones as well, you will find that they all fit in very harmoniously. The fifth is the next most concordant interval,

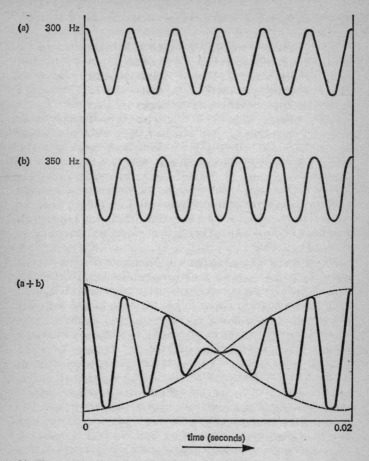

(a) 300 Hz

(b) 350 Hz

(a + b)

0   time (seconds)   0.02

11. The combination of sounds

being a fifty per cent rise in frequency over the lower note, because the difference tone falls at exactly one octave below the lower note.

A really crashing discord produces such a conflicting array of difference tones that there ceases to be any orderly relationship between any of the notes and there are all manner of throbs and wowing sounds. If you press the lowest two pedals on an organ with a sixteen-foot stop, you will get an excellent rendering of a marine diesel engine! Discord is really a matter of degree, and it is determined by the greater or lesser intensity of the beats that constitute the difference tones. Electronics engineers will already be conversant with the subject if they are familiar with the use of a heterodyne in radio signal transmission.

There is still a very common type of sound which has been left out, and deserves as much attention as the sound we have been discussing. All our musical instruments, and even the diesel engine, produce a sound whose waveform is such that it repeats itself periodically.

However, as many sound waves as are periodic are completely random or never repeat themselves at all. Listen to the wind, the rustle of leaves or the breaking of waves on the beach. Listen to the roar of an oil-fired boiler. They have no notes, no harmonics, no concord or discord.

All these sounds are the result of turbulence of some kind or another, and turbulence is simply random jostling, swirling and eddying of fluids. Pressure fluctuations are caused and these set up compression waves in the air in just the same way as any other sound source, but with no regular repetition or rhythmic movement. Random noise can also be caused by various other means, friction on rough surfaces, for instance.

Nevertheless, it does not follow that because this noise is irregular it cannot be defined in terms of frequency. A hiss is obviously higher pitched than a rumble, yet both are random.* This is because the pressure fluctuations in a hiss take place at a much faster rate than those in a rumble. Figure 12 helps to show this. Hisses and rumbles, though, sound completely different from periodic or harmonic sounds because the ear is not excited in the regular fashion which causes the sensation of distinct notes. One hiss will sound very much like another because there are no overtones or harmonics to give it character.

When it comes to dealing with noise from machines, random

* Strictly speaking the word 'random' is a misnomer as it should apply only to waves whose amplitude distribution as a function of time is Gaussian (see Glossary).

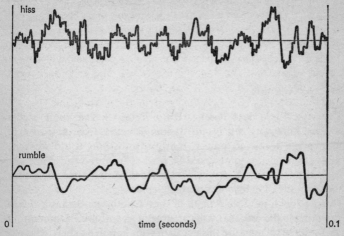

12. Waveforms of a hiss and a rumble

noise in fact usually forms the basis of the sound we hear. Pure
tones and harmonics are often added in, often in discordant combinations, and the result is a very complex waveform which contains periodic components superimposed on a random background.

4 Din in decibels

Everyone these days has heard the word 'decibel' and hardly anyone knows what it means. It sounds rather like an acoustical equivalent to 'candle-power' that has something to do with the sound of bells. Nothing could be further from the truth, and the only connexion with any bell is that it is named after Alexander Graham Bell, the inventor of the telephone.

A decibel is not solely a unit of sound, and is really not a *unit* of anything in the way that volts, metres or grams are. You could, if you wanted to, measure the length of your hair in decibels, whereas you would be hard pressed to express it in volts! If all this sounds odd, let me explain. You would not think it at all odd if I said that the distance from London to Inverness was twenty times as far as from my house to London. I could express any distance by comparing it with the distance from my house to London, say Piccadilly Circus. London to John o' Groats would be twenty-six times as far, Australia some 500 times as far away, but this does not mean that Australia is 500 units of anything away; these figures are merely ratios.

One of the measurable properties of sound is the amount of energy involved, and the intensity at any point can be measured in terms of energy flow per unit area, for instance in watts per square metre. The difficulties come when you start to write down the intensities of common noises. Take, for example, the quietest sound that the best of us can hear. The intensity would be about $0.000\,000\,000\,001$ watts/m^2. Now take one of the loudest sounds that one might encounter, not necessarily unscathed, a jet aircraft at 50 metres range. The intensity would be about 10 watts/m^2. At 100 metres from a Saturn rocket at lift-off, the sound intensity would go well over the 100,000,000 watt/m^2 level.

Quite obviously we would have terrible difficulty in coping with the enormous range of sound intensities, either by expressing them in units of energy or even as ratios. There is, however, a simple, if not obvious, way out of this difficulty. The intensity of the quietest audible sound was 0·000 000 000 001 watts/m². Now a mathematician would prefer to write this as 10^{-12} watts/m²; although this may look unfamiliar, remember that 10^2 means 'ten squared' or 100, and that 10^3 means 'ten cubed' or 1,000. In the same way,

10^{-2} means $\frac{1}{10^2}$ or $\frac{1}{100}$ or 0·01 and 10^{-3} means $\frac{1}{10^3}$ or 0·001. If you

multiply a number by 10^x, you are increasing it by a factor of ten, 'x' times.

In order to produce a manageable method of measuring sound intensities, let us start by expressing them as ratios, using 10^{-12} watts/m² as a reference, and noting the number of tenfold increases to which the reference intensity must be subjected for it to equal the intensity we are measuring. The noise of the jet aircraft is 10,000,000,000,000 (or 10^{13}) times as intense as the reference, or thirteen increases of tenfold. This cuts the immense range of intensities down to size and if we call a tenfold increase a Bel, we have a 'unit' in which to express the ratio. The intensity level of the jet aircraft noise is therefore 13 Bels. The Bel is rather a large unit, and so, to add a finishing touch, let us divide it into ten sub-units and call them decibels. The jet noise is now 130 decibels (abbreviated 130 dB) and to avoid confusion with any other reference intensity one can add 're 10^{-12} watts/m²'.

If the ratio of an intensity to the reference intensity is something less straightforward, like 8,300, it does not appear as though the conversion into decibels is going to be anything like as easy. We know that the number of tenfold increases is at least 3 and less than 4, but to work it out exactly would need some very lengthy calculations. How is this problem solved? The answer is that every ratio we are ever likely to use has already been worked out in terms of 'tenfold increases'. That is what logarithms are.

Any number can be expressed as ten to the power of something. 100 is 10 to the power of 2, and so 2·0 is the logarithm to the base 10 of the number 100. 3·0 is the logarithm to the base 10 of 1,000. Not so obviously 3·919 1 is the logarithm of 8,300. (There is no need to go on including 'to the base 10' because, as this is the most common base, it is assumed to be 10 unless any other is mentioned. Nevertheless, in formulae one should write '$\log_{10}$'.)

We can now rewrite the definition of the decibel as follows:

$$\text{Intensity Level (I.L.)} = 10 \log_{10} \frac{Intensity\ measured}{Reference\ intensity}$$

For example, take a sound intensity of 0·26 or 2·6 × 10⁻¹ watts/m². The intensity level in decibels re 10⁻¹² watts/m² is equal to:

$$10 \log_{10} \frac{2 \cdot 6 \times 10^{-1}}{10^{-12}}$$

$$= 10 \log_{10} 2 \cdot 6 \times 10^{11} \quad \text{dB}$$

The logarithm of 2·6 is 0·415, so the answer is 10 × 11·415 or 114 dB to the nearest decibel.

It is very important to remember that because decibels are not proper units like volts and ohms, one cannot treat them as such. If you put two 6-volt batteries in series, you have a potential difference across the two of them of 12 volts, but if you have a noise of 80 decibels and introduce another of 80 decibels, what do you get? 160 decibels? No, because if you double a number, its logarithm, to two places of decimals, will always go up by 0·3. Therefore, if you double a sound intensity, the I.L. will go up by 0·3 Bels, and thus 3 decibels.

This will hold good whatever the intensity level: *a doubling of sound intensity gives an increase of 3 decibels*. Table 1 shows the decibel increase for the addition of sounds of different intensity.

Now that the mystery of the decibel has been solved, it is worthwhile having a look at some more examples. Table 2 gives a table of typical noises and their intensities in decibels.

How do we find out the intensity of a sound? Well, it is very difficult, and much easier to measure the pressure fluctuations that the sound waves create. Table 3 gives a list of sound pressures alongside sound intensities. From this can be seen that the range of sound pressures is not as great as the range of sound intensities. Pressure in fact increases at half the rate that intensity does. When a sound pressure is doubled, it takes four times as much energy to increase the rate of oscillation of molecules (particle velocity) to achieve the doubling of pressure. Therefore, if we measure the sound pressure in decibels in the same way as we measured the intensity, and introduce a factor of 2, the answer should be the same, and is. The sound pressure of the quietest audible sound is about 0·000 02 Newtons/m², and of a diesel cab 2 Newtons/m², thus

$$20 \log_{10} \frac{2 \times 10^0}{2 \times 10^{-5}} = 10^5$$

$$= 100 \text{ decibels.}$$

Table 1. Addition of sound levels (to the nearest 0·5 dB)

Difference in dB Between Two Levels	Add to Higher Level
0	3
1	2·5
2	2
3	2
4	1·5
5	1
6	1
7	1
8	0·5
9	0·5
10	0

Table 2. Intensities of typical noises

Approximate sound pressure level dBA	Source and location
200	Moon rocket at lift-off, 300m away
160	Peak level at the ear, of 0·303 rifle
140	Jet aircraft taking off at 25m
120	Submarine engine room
100	Very noisy factory
90	Heavy diesel lorry at 7m
	Road drill at 7m (unsilenced)
80	Ice-cream van at 3m
	Ringing alarm clock at 1m
75	Inside railway carriage

Approximate sound pressure level	Source and location
70	Inside small saloon car, 50 k.p.h.
	3m from domestic vacuum cleaner
65	Busy general office with typewriters
	Normal conversation at 1m
40	Quiet office
35	Quiet bedroom
25	Still day in the country away from traffic

Table 3. Intensities, sound pressures and decibel levels for sound in air at room temperature and sea-level pressure

Intensity watts/m²	Sound Pressure Newtons/m²	Sound Level dB
100 000 000	200 000	200
10 000 000		190
1 000 000	20 000	180
100 000		170
10 000	2 000	160
1 000		150
100	200	140
10		130
1	20	120
0·1		110
0·01	2	100
0·001		90
0·000 1	0·2	80
0·000 01		70
0·000 001	0·02	60
0·000 000 1		50
0·000 000 01	0·002	40
0·000 000 001		30
0·000 000 000 1	0·000 2	20
0·000 000 000 01		10
0·000 000 000 001	0·000 02	0

The important thing to remember about sound pressure in decibels is that if you double the pressure you have a 6 dB increase. If the noise in the diesel cab rises to 106 dB, the sound pressure will have doubled to 4 Newtons/m^2 and the intensity will have gone up fourfold to 0·04 watts/m^2.

We have talked a lot about measurements of sound intensity, but have completely ignored the practical problem of obtaining the measurements.

A sound wave has several measurable properties. These include intensity, pressure, particle velocity and particle displacement. 'Particle' in acoustics is almost synonymous with 'molecule'. All of these properties are directly related, and therefore if any one can be measured all the others can be arrived at.

Many people will have noticed how lightweight surfaces can be seen or felt to vibrate when placed in the path of a sound wave. In fact the earliest type sound-level meter, an oscillograph, was a large diaphragm to the centre of which was connected a delicate thread, a mechanical system to amplify the vibrations, and a pen to record the displacement of the diaphragm on a roller, with a result that looked like the 'wavy lines' of the last chapter.

This system was hopelessly insensitive, and did little more than assist early scientists to prove their theories about sound. The inertia of the parts in the mechanism severely restricted the frequency response and accuracy. The inertia could be reduced by replacing the mechanical amplifier by an optical system and bringing in the help of photography. In the resulting device, the diaphragm's thread was wound around a drum, mounted on a spindle, to which was fixed a mirror to rotate with the drum. A beam of light was shone on the mirror, which was rotated to and fro by vibration of the diaphragm so that the beam was diverted and the displacement could be recorded on light-sensitive paper.

It was not until the advent of electronics that any sort of accurate measuring device was possible, and not until the availability of transistors was the modern hand-held sound-level meter able to be produced.

A modern sound-level meter is really an electronic analogy of the early mechanical version. The first stage of the process is the conversion of pressure fluctuations into voltage variations, and the device with which this is achieved is the microphone. There are many many types of microphone in use today, condenser, moving-coil, crystal, ribbon, hot-wire and Rochelle salt to name a few, and it is beyond the scope of the book to go into the workings of all of them.

They all carry out the same basic function and most consist of a diaphragm of sorts which is vibrated by the pressure variations in the sound wave. The displacement of the diaphragm causes the voltage across the microphone to vary in an identical fashion. The next stage in the measuring process is for the alternating current from the microphone to be rectified, and then amplified. The final stage is for the signal to be fed into a voltmeter, calibrated in decibels. In most instruments the voltmeter does not measure the peak values of the signal, but the 'root-mean-square' or R.M.S. value, which is a type of average, and of more direct use in most cases than peak values.

To cope with the enormous range of sound pressures, a straight-forward voltmeter is inadequate, and therefore the amplification stage of a sound-level meter contains a series of circuits ranged in 10 dB steps that can be progressively switched in.

A refined version of the early oscillograph is, however, still very useful today. The cathode ray oscilloscope has completely over-come the problems of inertia in the mechanical oscillograph because the electron beam has negligible mass and can be electro-magnetically deflected to show a graph of the voltage oscillations fed into the instrument. Apart from the fact that an oscilloscope trace can be used to carry out mathematical analysis of the wave-form, oscilloscopes are most useful when it is necessary to measure impulsive noises. An ordinary sound-level meter, we have seen, constantly takes an R.M.S. average of the signal. The noise from sonic boom or gunfire, though, is not so much a noise as a single pressure pulse of great magnitude and sometimes potential danger, followed by small pressure oscillations (figure 13). The initial jump in pressure could be quite enough to damage an ear or break a window, and because it occurs only once and for a very short time an R.M.S. value will be useless and very misleading. Although special impulse sound-level meters are becoming available, most meters would not even register the full R.M.S. value of the pulse because their response would not be quick enough. So the oscilloscope comes into its own, and shows instantaneously an accurate graph of the pressure rise so that the peak value can be measured from the screen.

13. Typical impulse noise

Throughout acoustics, the influence of frequency on the behaviour of sound is perhaps one of the most important factors. The human ear can pick up sounds as low as 30 Hz and at best as high as 18 KHz and therefore the sound-level meter must do the same. There is, however, one large snag. The next chapter will show that ears do not respond with anything like the same sensitivity to all frequencies and in fact a sound at 30 Hz has got to be 40 dB higher in sound pressure level to sound as loud as a sound at 1 KHz. Therefore the reading on a sound-level meter is never much use as it is.

Electronics engineers have applied their minds to this, with the result that modern meters have built in 'weighting' networks, which consist of circuits which can be brought into play to reduce the response of the meter to low-frequency and very high-frequency sounds and liken it more to an ear. There are three weighting networks, 'A', 'B' and 'C', but the 'A' network has been found to be the most useful, and the 'B' and 'C' are seldom used. Some modern meters have a 'D' network which gives values of 'PNdB' or perceived noise level (a unit described in the Glossary and most commonly used for assessing aircraft noise). When a sound-level meter is used, the results are therefore much more informative and comparable with loudness sensations if the 'A' scale is used, and these readings are called dBA. However, even dBA cannot convey more than a rough indication of the frequency make-up of a noise, and as this aspect is often so vital more equipment has usually to be called upon.

The frequency scale, like the intensity scale, is logarithmic, in that successive steps up the ladder represent doublings of the number of vibrations per second. The range, though, is not as great, and instead of counting the number of tenfold increases, or using logarithms to the base ten, actual frequencies are always quoted as the number of vibrations or cycles per second which make up the sound wave. These are called Hertz (Hz). An analysis of a sound into the intensity at each frequency would require an infinite number of readings. To get over this problem, the frequency range is divided up into octaves, derived from the musical equivalent. The highest frequency in an octave is twice the lowest. The first and simplest step in frequency analysis is to measure the sound pressure level in each of eight or eleven octaves, depending on the range required, by feeding the output of a sound-level meter into an octave filter set or octave-band analyser. The word 'band' is used to describe any section of the frequency spectrum. This instrument contains eight or eleven electronic filters, devices which will pass only the frequency components of a signal within their band, and by switching in one, the sound pressure level in the band is measured back on the sound-level meter.

Even octave-band analysers are not informative enough in many cases, and so more elaborate analysers have to be used: half octave or one-third octave units. For even more accuracy, narrow-band analysers, resolving the noise into either constant percentage bandwidth, say 6 per cent of the centre frequency or a 10 or 6 Hz bandwidth, can be used. When pure tones exist in the noise spectrum, as they often do, it is possible to determine their exact frequency and amplitude by using a discrete-frequency analyser.

These instruments are usually very bulky and are thus confined to the laboratory. The sound to be analysed is often recorded on a high-quality portable tape-recorder via the microphone and amplifier networks of the sound-level meter, using reference tones for calibration, and later replayed into the analyser, which will automatically write out the frequency spectrum on a roll of graph paper. Figure 14 shows a typical noise which has been analysed into octaves, one-third octaves and 6 Hz bandwidths.

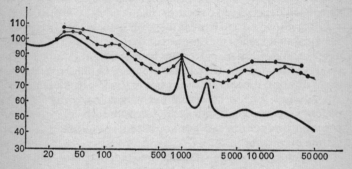

14. Analysis of a sound into octaves, one-third octaves and 6 Hz band-widths

Sound-analysing instruments are costly items – a full outfit including meter, tape-recorder, analyser and level recorder can cost upwards of £5,000. For this reason, and also because of the amount of detail involved in a frequency spectrum, sound levels are often talked about in terms of single units. The layman may well be confused by the array of units and systems that have been thought up to try to express the loudness or annoyance value of a noise. By far the most commonly used is the dBA, already explained, but there are also phons, sones, noys, noise rating numbers, PNdB used mostly for measuring aircraft noise, noise criteria, traffic noise index, and many more.

5   Hearing and deafening

Imagine a machine which could weigh just as accurately a flea or an elephant. The mind boggles, but the human ear can hear sounds the loudest of which can be over ten billion times as intense as the softest and analyse sounds over a range where the highest frequency is almost a thousand times as great as the lowest. One talks glibly about ears and hearing, but what is hearing? What goes on at the end of those waxy tubes? Without stopping to think the answer seems obvious: the ears are the organs of sound perception. They are, however, such delicate and intensely complex organs and perform in conjunction with each other with such computer-like precision that science still has not penetrated all the mysteries surrounding them.

What do ears do? They respond to sound, of course, and they analyse sound into frequency components, acting both as narrow-band analysers and discrete-frequency analysers. They can feed to the brain coded information of sufficient detail to be rich in meaning, whether it be the interpretation and identification of sounds or the understanding of complex speech. They serve as a 'feed-back' system to monitor their owner's speech; they are a direction and range finder, and are capable of fixing on to harmonic sounds in the midst of high-level random noise to make speech reasonably understandable in noisy surroundings. The organs of balance also form part of the ear. Although they look innocent enough, and their function is often taken for granted, ears are among the most complex of the organs of the body.

The ear is normally talked about as three separate sections: the outer ear, the middle ear and the inner ear (figure 15). The best way of looking at it is to follow the fortunes of a sound wave on its way to being heard. You may remember from chapter 3 that when a sound wave is travelling along a tube and comes to the open end,

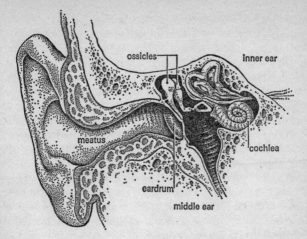

15. The ear

most of it is reflected because it suddenly meets an infinitely large body of air in the world outside. Exactly the reverse would happen if a sound wave travelling through the air met the open end of a tube: the sound wave which entered the tube would only have a fraction of the intensity of the wave outside.

The thing which is vulgarly known as the earhole and is more properly called the external auditory canal, or meatus, is of course a tube. On its own it would not be very successful at receiving sound from the outside world. In technical language, this is because the impedance of the meatus is a poor match to the impedance of the air outside, and, as in electronics, an impedance match is necessary for a good transfer of energy from one point to the next. The problem is solved by that flap of sinew and flesh known variously as the auricle, concha or pinna, which is sometimes ignorantly maligned and said to have no acoustic function. The auricle, with its curious shape, softens the abruptness of the change from free air to air in a tube as the sound wave meets the ear, and helps to 'funnel' more sound into the meatus. It is analogous to an aerial in electronics.

Once inside the meatus, the sound wave travels along as a plane wave for two to three centimetres until it meets the eardrum or tympanic membrane. This is merely a diaphragm kept under slight tension by muscles known as the tensor tympani and stapedius which is vibrated by the pressure fluctuations of a sound wave in the meatus. Attached to the eardrum are three tiny bones, the malleus, incus and stapes, or the hammer, anvil and stirrup. These are contained in an air-filled bony cavity. The stapes is shaped like a stirrup and the 'footplate' fits into the oval window, the entrance to the inner ear. The middle ear, containing these bones, is another impedance matching device, and is really an acoustic transformer. The three bones, collectively known as the ossicles, have a mechanical advantage of about three to one, and taking into account the much smaller surface area of the oval window compared with the eardrum, the pressure exerted on the oval window is about twenty times as great as the pressure exerted on the eardrum by the sound wave.

The middle ear is connected to the rear of the nasal cavity by the Eustachian tube, named after the Italian anatomist Eustachio. This tube opens on swallowing to allow the pressure on both sides of the eardrum to be equalized, for instance to compensate for altitude changes. You will no doubt at some time have had problems when travelling by air in a poorly pressurized aircraft with numbness or pain in the ear after landing or take-off, particularly if you had a head-cold at the time. The reason for this happening is that the atmospheric pressure on the outside of the eardrum rises or falls, but air pressure on the closed middle ear remains the same, with the result that a steady pressure is exerted on the eardrum and it is less able to vibrate, causing the slight deafness. If the pressure is great enough the result is earache.

Normally this state of affairs is very quickly put right by swallowing, thus opening the Eustachian tube, and allowing air to enter or leave the middle-ear cavity and equalize the pressure on both sides of the eardrum. The tube then closes again to allow the eardrum to resume vibrating. If you have a head-cold, the Eustachian tube and even the middle-ear cavity will fill up with mucus, and the pressure equalization could not take place. The middle ear often contains a little liquid in it which drains out during swallowing, but if it does fill up with mucus the viscous drag exerted on the ossicles reduces hearing sensitivity again.

The final stage in the hearing process occurs after the vibration in the ossicles has been transferred to the inner ear. The inner ear consists of a complex system of tubes embedded in the bone of the skull known as the labyrinth. This again is really two organs in one; part of it is the non-hearing vestibular system incorporating the semi-circular canals. This is a device for measuring angular acceleration and it thus enables the body to balance itself. The other part of the labyrinth is the cochlea, and this is the most complex part of the ear. It looks like a minute snail shell, and consists of a coiled-up tube filled with a liquid known as perilymph divided into parallel cavities, the upper gallery and the lower gallery, by the cochlear partition. The two galleries are connected at the end of the cochlea at a small gap called the helicotrema.

If the coiled-up tube could be straightened out it would look like figure 16(a). Movement of the stapes in the oval window causes the fluid to vibrate, this being possible because the round window at the end of the lower gallery has a flexible membrane across it which allows pressure in the near-incompressible fluid to be relieved.

The cochlear partition is formed by Reissner's membrane and the basilar membrane and contains another liquid, endolymph. In between these two membranes is the organ of Corti, which consists of about 24,000 fibres arranged rather like a row of harps. Experts are not in agreement about the exact method in which these fibres, which are connected to nerve endings and the auditory nerve, respond to vibration in the perilymph. Some say that the fibres are in tension and the resonant frequency (remember the stretched string) goes progressively higher with fibres farther away from the base of the membrane, and that frequency analysis is achieved because each fibre will only be excited at its own natural frequency.

The theory which now enjoys most acceptance is that originated by G. von Békésy. It suggests that the fibres are not in tension, but that complex mechanical forces result in the cochlear partition being deflected at different points along its length for different frequencies, and this causes the organ of Corti to be subjected to

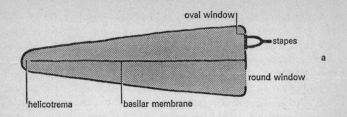

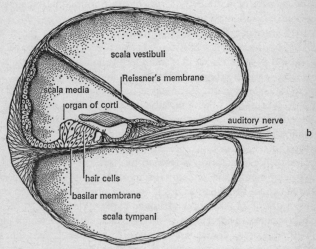

16. (a) A schematic diagram of the 'uncoiled' cochlea. (b) A section through a cochlear turn

shear stresses at these points, in turn causing excitation of those particular nerve fibres.

So much for physiology. Let us go on to the subject of what the ear hears, which is of more relevance to the noise sufferer than how it hears. The lowest pitch which the average normal person can hear has a frequency of about 20 Hz. It is rather difficult to draw the line at the bottom end of the scale because lower-frequency sounds than this can often be perceived, not by the organ of hearing, but by other parts of the nervous system. The top end of the scale varies immensely from person to person. Of particular importance is the age of the listener, and a person of eighteen with perfect hearing might just be able to hear up to 20,000 Hz (abbreviated 20 KHz for 20 Kilohertz) but usually 18 Hz to 16 KHz is about the limit at any age. Some people have faint singing in their ears at very high frequency most of the time which makes it difficult to distinguish an external stimulus at these frequencies.

The words 'pitch' and 'frequency' cannot really be interchanged. Frequency is a physical quantity of a sound wave, and pitch is a subjective sensation which is not absolutely dependent on frequency. At low and high frequencies the pitch of a note tends to rise at high intensities, and at low intensities at low frequencies the note can be varied in frequency by 5 per cent or so without the listener noticing.

Another reason why pitch and frequency are not synonymous is that often complex sounds possess a particular pitch and in fact are made up of a number of frequencies. A note from a violin will have one definite pitch but will contain many harmonics of differing frequencies.

The pitch assigned to such a note is usually that of the fundamental tone or first harmonic, but it is even possible to remove the fundamental from a note and the pitch will not be heard to alter; only the tone colour or timbre would change. This is really a sort of aural illusion, because it can be shown that in such cases the cochlea is not excited in the area normally associated with that frequency.

As a person gets older, a completely normal process known as presbycousis occurs, which progressively causes the sensitivity of the ears to high-frequency sound to fall. Figure 17 shows the normal course of this process. This graph, which shows the lowest sound levels which are audible at each frequency, shows up one of the most important facts in the whole of acoustics – that the ear is very much more sensitive to sound around the 4 KHz region than to low- or high-frequency sound. A sound at 30 Hz cannot normally be heard until its sound pressure level reaches nearly 60 dB, whereas at 4 KHz the threshold of audibility, as it is known, for a young person with good hearing can be as low as −2 dB. Do not be confused by the minus sign; it does not mean that there is 'less than no sound'. Remember that sound pressure levels are normally calculated in decibels with reference to 0·00002 Newtons/m². If the sound pressure to be measured is lower than this the level in decibels will have a negative sign. In practice, there would never naturally be surroundings where the background noise level was low enough to enable a subject to hear a sound of −2 dB. This difference in sensitivity becomes slightly less marked as the level of the sound increases, as figure 17 shows, and if the sound pressure level reaches much over 130 dB at any frequency the effect is painful. Instant damage to the ears would occur if they were exposed to sound levels in the region of 150 dB at any frequency.

We will come to the subject of deafening by noise later, but at this stage, before the physiology of the ear has been forgotten, it is interesting to know about the two safety mechanisms in the ear which help to protect it against damage. The first is known as the aural reflex. You will remember that the eardrum is kept under

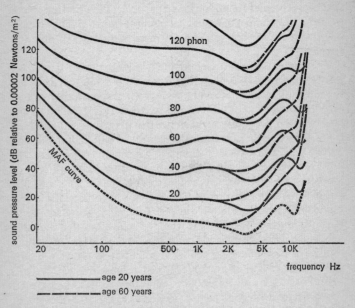

age 20 years

age 60 years

17. Equal-loudness contours for pure tones

slight tension by a muscle called the tensor tympani, really consisting of two small muscles which can draw the eardrum inwards and the stapes outwards. If the ear is subjected to a noise of more than about 90 dB which lasts longer than 10 milliseconds or so, a reflex action will occur which will tighten the tensor tympani and stiffen up all the mechanical parts in the middle ear, and the result is a reduction in sensitivity to low- and middle-frequency sound. Some lucky people have voluntary control over the aural reflex, and if they have warning of an impending loud noise they can activate it and get immediate protection.

The second protective device concerns the mode of oscillation of the ossicles. In normal operation, with normal-intensity sounds, the ossicles vibrate in such a way as to push and pull the stapes and thus via the footplate in the oval window push and pull on the perilymph in the upper gallery of the cochlea. If, however, you are unfortunate enough ever to be subjected, bare-eared, to a noise of over 140 dB, the whole mode of oscillation of the malleus, incus and stapes changes and they rock from side to side. Thus the footplate is no longer pushed and pulled, but rocks to and fro and pushes the perilymph from side to side of the oval window instead of up and down the cochlea. As a result the pressure fluctuations in the perilymph are greatly reduced, and the effect is a sudden drop in loudness as the second mode of oscillation comes into action.

Looking again at figure 17 one can see that not only is the ear very much less sensitive to low-frequency sound, but it is also non-linear in its assessment of loudness. You might think that if you double the sound pressure you will double the loudness, but this is not so. If you remember reading the previous chapter, you will know that a doubling of sound pressure gives an increase of 6 dB. However, if you listen to a sound, measure it in decibels, and then increase its volume until you consider it is double the loudness, you will find that the increase in sound pressure level is about 10 dB; in other words the sound pressure has more than trebled. Again from chapter 2, we know that this means that the intensity of the sound (the rate of energy flow per unit area) has gone up tenfold.

The ear as a sound analyser has a major 'failing'. This is that the presence of, say, a single note of a certain frequency renders notes of frequency close to it and of lower intensity inaudible (figure 18). This effect is known as masking, and occurs with all types of sound beside single tones. It is, however, often a great advantage. For instance when it is required to achieve speech privacy it is often easier to introduce a masking noise rather than reduce the sound level of the speech.

What about the other properties of our hearing mechanism, those of direction and range-finding, and of identifying single

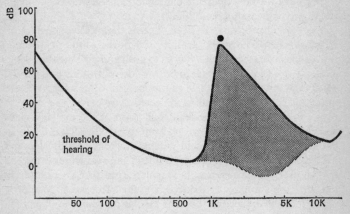

18. The masking effect of a pure tone

components amid a whole spectrum of noise, as opposed to merely sensing the pitch of a note? Both these facilities are made possible by the fact that we have two ears and are largely denied to people who have poor hearing in one ear. The determination of direction is best explained by looking at figure 19. Except when a sound source is situated on the axis of the head, the distance from the source to each ear will be different, with the result that it will take a tiny fraction of a second longer to reach one ear than the other. The brain is able to 'measure' this time difference (figure 19) and assess the direction of the origin. This alone will not be very accurate because unless the range of the source is known the angle off the axis cannot be calculated merely from knowledge of the time difference. However, from experience the listener can often

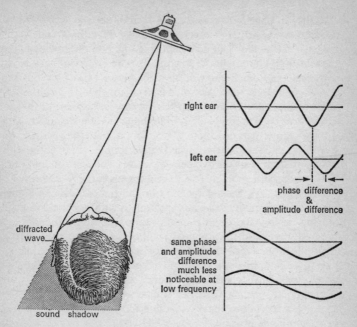

right ear

left ear

phase difference
&
amplitude difference

diffracted
wave

sound shadow

same phase
and amplitude
difference
much less
noticeable at
low frequency

19. Identifying the direction of a sound

tell the range of a source from its loudness, and for distant sources
from the frequency spectrum which will be altered by atmospheric
absorption and the effect of the environment, tending to damp out
the high-frequency components. Also, except for sources nearly
on the axis of the head, one ear will always be 'round the corner' or
in a sound shadow and the sound which reaches it will have diffracted round the face or the back of the head. This accentuates the
difference in the signals from the two ears.

As a result of all these factors, it is easy to understand that it is
more difficult to locate low-frequency sources than high-frequency
ones because the time difference is much less obvious (figure 19).
In addition, we shall see from chapter 6 that low-frequency sound
is diffracted round corners to a much greater extent than high-
frequency sound, reducing the difference in reception at the two
ears.

The other benefit of bin-aural hearing is bound up with the basic difference between harmonic sound made up of pure tone components and random sound which is just an aimless jostling of air molecules. If you listen to a harmonic sound amid wind noise, for instance, the human ear will be a lot better at picking out the former sound than any electronic instrument short of a computer. The reason is fairly simple and the process is the same as that of correlation used in sound analysis with the aid of a computer.

Most adults' ears are about 150mm apart, with the result, as we have said, that they nearly always receive sounds out of phase with one another. If you take two periodic waves which are out of phase and add or, better still, multiply them together, the resultant wave will have the same basic shape but with much increased amplitude. If you take a random wave and multiply it with a replica which is lagging behind, the result is again random, but nothing like the original wave, and the amplitude will be barely increased by the multiplication. This is how it is possible to extract a periodic or harmonic sound from a random background. The last few paragraphs will have special significance to hi-fi enthusiasts with stereophonic systems, which of course were invented to give the sound they reproduce, that extra 'dimension' gained with bin-aural or stereophonic hearing.

Without straying too far from the subject of this chapter, it is interesting to look at one more phenomenon of bin-aural hearing: ventriloquy. Ventriloquy, in the flesh, is a much more complex illusion than it seems and there is more in 'voice-throwing' than you might think. We have seen that the weak link in the sensitivity of two ears to direction (which can be remedied with electronic instruments by the use of a third 'ear' or microphone) is that the time or phase difference at each ear indicates direction only if the range is known, and judgement of this is sometimes vague. If someone is speaking to you, it is rather obvious where the sound is coming from because people do not often flap their lips and jaws for any other reason. As long as the origin of the sound is obvious, it will not be possible to fool the brain into thinking that the sound is coming from anywhere else. But if the speaker clenches his jaw and speaks without any visual sign, it is no longer possible to determine the range of the source visually, and the direction-finding process becomes vaguer. If the ventriloquist is skilful enough to engage the interest and fascination of the listener, and can by non-acoustical means focus the listener's attention strongly enough on another, not too distant object, the brain will take the visual distance of that object into its calculations of phase difference at the ear, and the result can be a genuine sensation that the sound is coming from that object. If the object is an attractive and well-animated dummy, the effect can be extremely good. However, nowadays ventriloquists are usually seen on television, so that the acoustic side of ventriloquy is lost, reducing the act to a puppet show and an exercise in keeping the mouth still.

Ears, of course, do not always work as marvellously as they should. Three things can severely affect their workings: accident, disease and noise. Eardrums can be ruptured by foreign bodies, and part of the middle or inner ear can be damaged by a blow on the head. Disease can affect the middle ear, or can eat away the sensitive nerve-hairs in the basilar membrane. Worst of all, the auditory nerve and its connexions in the brain can break down, and the unfortunate person becomes 'stone-deaf' or nerve deaf.

All but the last type of ear disease can be helped by medical science. Eardrums and ossicles can be transplanted or replaced with plastic substitutes, and when the nerve-hairs in the cochlea start to lose sensitivity, amplification of the sound entering via the meatus can help; but when the auditory nerve goes, the ear as a sense organ becomes useless. I am sure that it will eventually become possible to stimulate other parts of the nervous system in such a way as to produce an artificial sense of 'hearing'. The sensation of pain has much in common with the sensation of hearing. Pain often has 'pitch' characteristics: a pin prick is high-pitched, a headache low-pitched. It may before long be possible to use this part of the nervous system to enable a nerve-deaf person to perceive sound. It is certainly possible to give a blind person a very basic form of 'sight' by treating the back as a sort of television screen and covering it with pressure pads which are activated in accordance with solid objects in front of the subject. Further research is in progress regarding the direct injection of signals into the brain itself.

The ear may suffer from disease, but in many ways it is much

more serious that very few adult men have hearing that is un-
damaged as a result of noise. It is therefore nearly impossible to
establish what 'normal' hearing is for men. Why is this? There are
a number of reasons. Few men reach middle age without having
used firearms with unprotected ears; many have seen active
service. Although all the services are now fully awake to the
dangers of noise, and in fact make valuable contributions to our
knowledge of the subject, many amateur marksmen and sportsmen
are oblivious to the fact that every cartridge fired means another
step down the ladder to severe hearing loss. Schoolboys are even
subjected to gunfire in cadet corps which is a serious hearing
hazard, particularly in indoor ranges, and the 12-bore shotgun
can do just as much harm. Ear protection, which is discussed in
chapter 14, should be mandatory for everyone who uses any gun,
although great care must also be taken that this does not lead to
accidents from inability to hear shouted warnings.

The most widespread and serious cause of noise-induced
hearing loss is subjection to high noise levels in the subject's place
of work. This may be the cab of a diesel truck, an iron foundry, or
in a host of factories in trades as diverse as printing and plastics.
Apart from gunfire and explosions, the likelihood of a person
sustaining hearing damage from noise other than at work is
remote. However annoying aircraft noise or traffic noise may be,
it is very unlikely that the victim will suffer physiological damage
of any sort. The exception perhaps is in the case of the riders of
some motor cycles, and it has been said that pop-music can be
harmful.

In view of all this, perhaps we should take a close look at what
noise actually does to its victims. How much noise is dangerous?
Is the damage reversible?

Noise can affect ears in three ways. It can deafen or damage ears
instantaneously; it can severely reduce the ear's sensitivity to
sounds of certain frequencies over a period of time; it can numb
the ears for a limited period of time, and return to near normal
within a matter of minutes, weeks or months.

The first sort of damage, acoustic trauma, is usually the result
of a very high-intensity impulse noise, perhaps from an explosion.
For obvious reasons, it has not been possible to determine by

experiment the level of noise required to do this, but an impulsive noise over 150 dB would probably be instantly damaging. The eardrum can be ruptured beyond repair, and the ossicles can be broken or displaced. The cochlea, however, would probably survive, because failure of the ossicles would prevent the full force of the noise being transmitted into the perilymph fluid.

While on the subject of impulsive noise, we should notice an important point arising out of the previous chapter. Most sound-level meters measure not the peak value of a sound wave, but the root-mean-square value, a type of average. This is fine for continuous noise, and it gives results which tie in well with subjective aural assessment, but when it comes to an impulsive noise, there will often be only one single peak, whose rise and decay will be very sharp, with the result that an R.M.S. reading would hopelessly underestimate the level of the impulse. As well as this, few ordinary meters respond fast enough for the indicating needle to get up even to the R.M.S. value. The best way of measuring impulsive noise is either to show the waveform on a cathode-ray oscilloscope and measure the peak level from graduations on the screen, or to use a specially designed impulse sound-level meter.

Explosions are not the only source of impulse noise. If you strike a piece of steel plate with a large hammer, you will produce a sizeable impulse, but not of the level of an explosion. Lower-intensity impulses still cause damage, not to the middle ear, but to the inner ear in the same way as continuous noise, which we will go into shortly. It is very interesting to see that in this respect man has outstripped the evolutionary process. We have seen how the ear has two built-in protective systems, one of which is the aural reflex. This, unfortunately, takes 10 milliseconds or so to come into action, by which time the damage from impulsive noise can have been done. Now this type of impulse noise with a very short rise-time almost never occurs in nature, and is an exclusively man-made phenomenon. So nature has not really 'slipped up' in allowing the delay in the aural reflex, but has understandably not bargained for something which barely existed for millions of years. The next step in the evolutionary process in this respect will undoubtedly be to give more than the present élite the ability to activate the aural reflex consciously when there are signs of an impending loud noise. I think it may be possible to learn to stimulate this reflex. It is not always difficult to detect it coming into action, and occasionally there is an audible click, no doubt due to the presence of wax near the eardrum which is displaced as the eardrum tenses up. This is followed by a drop in hearing sensitivity.

Another major source of impulse noise is, of course, sonic boom. However, let me first say that it is generally believed that it would require a peak overpressure of 35,000 Newtons/m^2 to rupture an eardrum and 100,000 Newtons/m^2 to damage a lung. Overpressure from supersonic airliners would very seldom exceed 100 Newtons/m^2. People have experienced no physiological ill effects from overpressures over 6,000 Newtons/m^2, so there is a large margin of safety. The psychological reaction becomes significant, and public reaction audible, at about 50 Newtons/m^2, and between this value and 100 Newtons/m^2 rare minor damage to buildings occurs, but public reaction becomes vociferous.

You may be wondering at this stage why this type of noise is not expressed in decibels like all other noise. It certainly can be: the figures just quoted work out at 185 dB for eardrum rupture, 194 dB for lung damage, 134 dB for vociferous public reaction and 128 dB for some public reaction, all re 2×10^{-5} Newtons/m^2. However, sonic boom sounds so different from ordinary noise, and is so sudden, that decibel figures would tend to confuse matters. In addition, when one is working out the effect on people and structures, one needs to know the actual peak pressure exerted by the wave. We shall be taking a look at the causes of sonic boom in chapter 6.

Ear damage from impulsive noise is not, however, the main cause for concern. Far more prevalent are the effects of continuous periods of high-intensity noise. This affects people in two ways, the first of which is not necessarily very harmful. If you are exposed to sound pressure levels in excess of 90 dB or so in the middle- to high-frequency range for upwards of a few minutes, you will afterwards suffer what is known as a 'temporary threshold shift'. The normal threshold of hearing is the lowest level at which you can hear a sound of a particular frequency, and after exposure to a loud noise this rises considerably. For instance, you might normally be able to hear a 4,000 Hz tone at a sound pressure level of 5 dB. Usually, in fact, the background noise level is much higher than this and hearing measurements have to be undertaken in specially constructed rooms with a very low ambient noise level, using headphones to transmit the test sound. This technique is called audiometry, and the result is a graph of the subject's hearing sensitivity known as an audiogram. Audiograms usually show the deviation from normal sensitivity rather than the actual hearing threshold.

If you were exposed to a band of noise in the 1,200–2,400 Hz range having a sound pressure level of, say, 100 dB for ten minutes, you would find that immediately afterwards your hearing sensitivity would have dropped. Whereas you could hear the 4,000 Hz tone at a level of 5 dB before, after exposure to the noise the tone would have to be raised to 20 dB before you could hear it. Your threshold of hearing at 4,000 Hz would have gone up from 5 dB to 20 dB, but this loss would wear off in about half an hour's time, after which any permanent threshold shift which might have resulted from this exposure would be too small to measure. Figure 20 shows an audiogram of a typical person's hearing immediately after the noise exposure just described.

You may have noticed that there was a discrepancy in the frequencies in this example. The threshold shift was most marked at 4,000 Hz, but there was no element at this frequency in the noise which caused it. This is a basic characteristic of noise-induced hearing loss and the reasons for it are not fully understood. In nearly every case, though, the greatest threshold shift occurs at a higher frequency than that of the noise that did the damage.

As the exposure times get longer and the level of the noise higher, so the temporary threshold shift and the recovery time increase. If you were exposed to the 100 dB noise at 1,200–2,400 Hz for ten times as long, that is 100 minutes, you would have a temporary threshold shift of over 30 dB, and it would take about 36 hours for you to recover normal hearing.

Providing one is not exposed to these bursts of noise regularly, their permanent effects are so small as to be negligible. However, in countless factories throughout the world and in workplaces other than in factories, people are constantly being subjected to high noise levels; the effects cease to be temporary, and over a period of years become severe and chronic. It is a common oc-

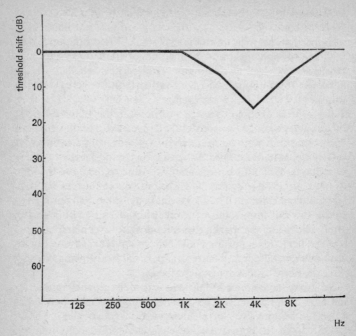

20. An audiogram showing typical hearing loss after a short period of exposure to noise

currence for a noise victim to deny that there is anything wrong with his hearing. I have been in many noisy factories and been told that they agreed there was a high noise level, but that they got used to it and their 'tolerance' increased. This is absolute nonsense. What happens is this: a man arrives to work in a noisy environment, and is certainly unused to the noise. At the end of the first day he will have a hefty temporary threshold shift, probably accompanied by ringing in the ears, known as tinnitus. If he is driving home, he will notice that his car engine sounds much more luxurious because he cannot hear the rattles and squeaks, particularly those whose frequency is around 4,000 Hz. When he meets his wife, her voice will sound just as loud, but as though she were speaking through a blanket. Unless he is subject to other hearing disorders, he may notice that very high-frequency sounds seem unnaturally enhanced. This is because the loss of sensitivity around 4,000 Hz contrasts markedly with the negligible loss at higher frequencies. The noise will also have had psychological effects on him and made him very tired.

By the morning, his hearing will have partly recovered, the tinnitus will have stopped, and he will have slept off the tiredness. The second and subsequent days will not have nearly such an effect on him. He will go to work with a threshold shift, and consequently the noise will not sound so bad. He will become accustomed to having a degree of hearing loss, and will probably have ceased to experience the ringing in the ears. Depending on the type and level of noise, for a considerable time – up to a year or so – he could be taken from his noisy job and given a quiet one, and after a few weeks, or even months, his hearing would return to within tolerable limits. However, there comes a point of no return, and eventually he will start to notice his less acute hearing, and speech will become less intelligible. He will find that he can only hear the television well if it is loud, even to him.

At this stage the damage becomes permanent and irreversible.

As the years go by the dip in his audiogram at 4,000 Hz will deepen and then widen to drag down the other frequencies with it. He will reach old age very deaf indeed (figure 21).

So much for the symptoms. What about the causes? How loud must noise be to do damage, and what kind of noise? A great deal of research has been done on this subject, and it is now possible to say with some authority the limits beyond which no one should be subjected. As a general rule, if the noise is so high that conversation is impossible and people are subjected to it eight hours a day, five days a week, then there is a danger that they will sustain permanent noise-induced hearing loss if the exposure continues over a long period.

The first thing one must ascertain is the nature of the noise. Is it random, broad-band or continuous-spectrum noise? In other words, does it have components of distinct frequency, such as pure tones, or is it merely a loud hiss or roar covering a wide range of frequencies? If it is the latter, and continuous, then Table 4 can be taken as showing the levels in dBA for one exposure per day,

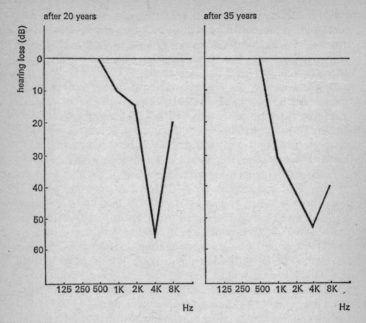

21. Typical hearing loss from weaving

which if not exceeded for longer than the stated periods indicate that the existence of a hazard to hearing is unlikely. If it is clear that the noise is very evenly distributed over the spectrum and not concentrated within a particular octave band, then these figures may be exceeded by 5 dB. (No one should ever be exposed to a noise level as high as 135 dB for any period, however short, and even with ear protection the absolute limit is 150 dB.) These figures are based on the criterion that a sound environment is regarded as acceptable if it produces, on average, a permanent noise-induced hearing loss in people after ten years or more of near daily exposure, of not more than 10 dB at 1,000 Hz and below, or 15 dB at 2,000 Hz, or 20 dB at 3,000 Hz and above.

If, however, the noise contains pure tones, the limits of acceptability are lower, and are shown in Table 5. The reason why pure tones

Table 4. Damage risk criteria. One exposure per day to noise having most of its intensity spread over about one octave. For very broad-band noise, the limits may be raised by 5 dB.

Duration	Limit, dBA
1½ minutes or less	120
3 min.	110
7 min.	103
15 min.	97
30 min.	93
1 hour	90
2 hour	87
4 hour	85
8 hour	85

Table 5. Damage risk criteria. One exposure per day to pure tones or narrow-band noise.

Duration	Limit, dBA
1½ minutes or less	115
3 min.	105
7 min.	98
15 min.	92
30 min.	88
1 hour	85
2 hour	82
4 hour	80
8 hour	80

are more damaging than broad-band noise is thought to be that they excite a section of the basilar membrane in the cochlea of the inner ear into resonance. You will have heard of opera singers purportedly breaking glasses by striking high notes, which happens because the glass is excited at its resonant frequency and the amplified movement of the sides eventually sets up stresses sufficient to break the glass. Now a band of random noise such as a hiss which contained the resonant frequency of the glass, however loud, would not normally break the glass. It would start the glass 'singing' but there would be no great amplification because of the lack of coherence in the noise reaching it.

It is possible to go a great deal further on the subject of damage risk specifications. For instance, if the noise is intermittent or of changing frequency, higher levels can be tolerated, but in these cases assessment of a likely hearing hazard must be made by a specialist.

Noise-induced hearing loss, or occupational hearing loss, is perhaps the most serious effect of noise, but it is by no means the only effect. There are many other ways in which it affects human beings: certain types of noise and vibration cause disease; noise can seriously impair communications; it can cause accidents; it can cause psychological disturbance through persistent annoyance; it can disrupt or prevent sleep, with serious results. In general, noise lowers the quality of life.

The pathological aspect of noise and its accomplice, vibration, is not yet fully understood. Users of hand-held vibrating tools are known to suffer injuries which have names like 'white fingers', 'dead-hand', 'pneumatic drill disease' and 'Reynaud's phenomenon'. The symptoms are pain, numbness and cyanosis of the fingers exposed to cold. Quite often there is damage to the bones and joints in the hand, with joint swelling and stiffening. It is possible that the bone and joint damage is the result of the repeated sharp blows the hands receive from hammering machines, and that the other symptoms are caused by high-frequency vibration. I have a machine as innocent as a car vacuum cleaner which is hand-held and vibrates at 290 Hz. It produces numbness and faint tingling in the hand after about half an hour's use. I have also found that vibration of the hand can cause temporary threshold shift in the hearing, by being transmitted via the bones of the arm and neck into the inner ear. The offending machine was a hand-held vibratory engraving tool, vibrating at 50 Hz, and after using it for half an hour, wearing ear defenders to eliminate any possibility of threshold shift due to airborne noise, I sustained the equivalent temporary threshold shift that would occur after exposure to an octave or so of high-frequency noise at about 90 dB for the same period of time. If such a tool can produce temporary threshold shift it is certainly capable of inflicting permanent damage on long-term users. It is nearly always the harmonics of the machine frequency which have the most harmful effects, and these exist even in a pneumatic drill up to several Kilohertz. This causes damage to the peripheral nerves and to the capillary blood vessels in the fingers and hand.

The other injurious effects of noise and vibration on the body are not at present thought to be serious, except for very low-frequency or very high-frequency sound and for very high intensities. High-intensity noise can cause resonance in the semi-circular canals, which are the organs of balance in the inner ear, with subsequent feelings of dizziness and nausea. Ultrasonic noise, of frequency above the audible range, can also cause nausea, and infrasonic noise and very low-frequency audible noise can excite resonances in organs of the body, including the heart and lungs. It may be possible to stop a heart from beating by acoustic excitation at the right frequency and of sufficient amplitude. Breathing can also become difficult in the presence of intense low-frequency noise.

At middle frequencies in the audible range resonances in the bones of the face and head make exposure to high intensities, even with ear protection, very unpleasant. It becomes difficult to think clearly, possibly because of resonances in the brain. Coordination of the limbs can break down, which may be due to brain or nerve vibration. If for any reason a person simply has to be exposed to very high-intensity noise, he must at least wear a sound-resistant helmet and not just ear protection.

The psychological and other non-pathological effects of noise are also very important, but sometimes extremely difficult to measure. How do you measure bad temper? How much damage is done by bad temper? People can become unnaturally violent, or may make very wrong decisions with disastrous or expensive results. The victims may become depressed and succumb to psychosomatic diseases; marriages and homes can be broken; accidents can be caused; industrial relations can be strained.

Then there is mere fatigue and the inability to concentrate. The result is inefficiency and more accidents. The relationship between efficiency and noise is again difficult to measure, because if you take a group of subjects and experiment with any aspect of their environment, be it acoustics, lighting or heating, their productivity and efficiency will improve merely because they feel that someone cares about their welfare and is bothering to do something about it. Few people would, however, disagree that people working in a high noise level make more mistakes and are thus less productive and efficient. It has also been found that absenteeism is reduced if the noise level is cut down.

Finally, there is the complex subject of communications. In many walks of life it is vital for one person to be able to pass information quickly and accurately to another. For thousands of years one principal means of communication has been the spoken word, and although the advent of electronics has reduced this dependence, speech is still essential to us. If communications are disrupted, two things can result. The first is once again inefficiency, but there is a more serious result which can be fatal. Frequently possible fatal accidents are avoided by someone shouting a warning, whether it be 'look out' as you are about to fall into a hole or a more complicated warning such as 'those wires are live'. Obviously if these warnings cannot be heard above the noise in the area, people will die for reasons which could have been prevented.

6 Why so noisy? – machines and noise

It is obvious to everyone that the noise problem of our age is not due to pistons in tubes or pulsating balloons. Nor can it be blamed on recorders, violins and cymbals. Yet these are the only sources of sound that we have so far talked about. Who are the real culprits? What is it that makes them noisy? How do recorders, violins and cymbals come into it at all?

Let us first take a look at machines in general. Thinking back to the musical instruments, we remember that there were three categories of instrument, those that produced their sound for aerodynamic reasons, those whose sound was caused by sustained mechanical vibration and those that involved impacts or percussion. There were also three stages in the production of sound: the initial disturbance, the modification or amplification of the disturbance and the radiation of sound. One important point arose, that without the initial disturbance no sound at all could be created. In the recorder it was vortex formation around a jet of air, in the violin it was friction between the bow and the string and in the cymbal it was the sudden bending or distorting of the cymbal as a result of a blow from a solid object. What about machines? What are the initial disturbances in them?

Take a simple machine with no moving parts, a blow lamp. It is an example of a purely aerodynamic sound source. Paraffin under pressure is vaporized and after expulsion through a tiny orifice emerges as a jet of inflammable vapour. This is continuously ignited, and gases, the products of combustion, are accelerated out of the nozzle. Noise is produced in three ways, firstly as the jet of vapour emits from the orifice exactly the same sort of vortex formation as occurs with the jet of air from the recorder mouthpiece. The result is a hiss. Ignition occurs a short distance from the jet orifice after air from outside has been sucked into the barrel of the blow lamp through large holes around the jet. The fuel and air mixture is already turbulent and so combustion is not a steady process. Pressure from expanding products of combustion is by no means smooth and steady and combustion noise is created. If you were to measure the fluctuations of light output from a flame like this, and compare them with the fluctuations in air pressure, in other words sound, you would find a remarkably similar pattern. Still more noise is created by the fact that at the end of the barrel there is a fast-moving turbulent stream of hot gas which suddenly has to mix with the surrounding still, cold air. There is a consequent swirling and eddying for some distance from the blow lamp, causing random pressure fluctuations and adding to the roar of combustion.

This is one type of aerodynamic noise, but almost more common is the straightforward explosion. Instead of a continuous controlled process of igniting inflammable fuel, there is a sudden combustion of a large amount of highly inflammable substance, bringing about a sharp rise in pressure and much turbulence. Whether the noise which results is a boom or a sharp crack depends on factors like the rate of combustion and whether it takes place in a confined space, such as a cartridge, when pent-up pressure is suddenly released as the bullet leaves the case. The sudden release of compressed gas or air can also be caused without the help of explosions, as for instance in the exhaust port of a pneumatic drill.

These are not the only types of initial disturbance caused aerodynamically. Whenever air moves past a solid object or vice-versa there is vortex formation, even in some cases with a streamlined object such as an aerofoil. Figure 22 shows what happens. Take the case of a simple sphere moving through static air: the sudden reduction in pressure in its wake causes turbulence, and thus noise. The better the aerodynamics of the object, the lower the noise. However, not only does the moving body create turbulence in its wake, but it also has to push the air aside in order to get through it. This means a sudden rise in pressure on both sides of the object, or, in the case of an aerofoil, a rise on one side and a fall on the other, and like any other pressure change this constitutes a sound wave and travels away from the object continuously at the speed of sound. If a regular series of objects passes a point in air at an even

22. Vortex shedding by circular section

speed, a short distance to the side of that point an observer will receive the effect of each object pushing the air aside one after another. If the objects are moving fast enough, the chain of pressure pulses will be audible sound. The obvious occasion when this occurs is when a propeller rotates. As each aerofoil blade passes any particular point it pushes or pulls the air around it and the successive pushes or pulls travel outwards as sound waves. If the propeller has four blades and rotates at 1,200 r.p.m. or 20 revolutions per second, for instance, a blade will pass our selected point 80 times a second, causing a fundamental note of frequency 80 Hz.

The pressure pulses from the blades will not be neat pure tones or sine waves. In chapter 2 we came across Fourier's theorem, which was that 'Every finite and continuous periodic motion can be analysed as a simple series of sine waves of suitable phases and amplitudes.' Now the pressure pulses from the blades are finite, continuous and periodic, and are thus capable of being analysed into Fourier components. If this is done, we find that the blade noise is in fact made up of a series of harmonics.

The whole effect is accentuated if a stationary object is placed very close to the rotating blades. The first thing that happens is that the air which is being swirled around by the blades moves past the stationary object, causing the same effect as if the object itself were moving. In addition, if you were to put your finger between the blades and the stationary object it would either be seriously crushed or be severed altogether. In just the same way, the air between the rotating and stationary parts is squashed. The compression of the air once again causes a sound wave. A third effect caused by the introduction of the stationary components is that as the rotating blade passes between the stators, it experiences a changing air velocity because obviously immediately behind a stator member there will be little air flow, whereas between the members it is moving with a fairly high velocity. Changing velocity means changing pressure, which in turn means sound. The frequencies of the noise from the rotor/stator interaction are dependent on the number of blades and relative speed. These principles apply not only to air, but also to liquids, and one can often substitute 'hydrodynamic' for 'aerodynamic'.

What about sustained mechanical vibration? There is a big difference here. With the aerodynamic sources, although we have only been talking about the initial disturbance, not the amplification or radiation of noise, we shall soon see that with aerodynamic noise all three components are often combined in one process. Not so with mechanical vibration, where the radiation part of things is often inefficient (thankfully). The simplest type of initial disturbance involving sustained vibration occurs when you draw an object across a rough surface. The noise of pencil on paper is a tame example, that of a chisel on a grinding wheel less so. The process is not difficult to understand. The surface of the paper on a small scale, the surface of the grinding wheel on a larger one, consist of random undulations. When the pencil or chisel passes over the surface it is vibrated up and down by the undulations of the surface.

Another type of vibration, unlike the random one just described, is caused by the rotation of unbalanced components, causing periodic displacement of the machine. The simplest case is that of a weight on the end of a rotating arm. Centrifugal force acts in whichever direction the arm is pointing at the moment of interest, and if the arm rotates with uniform velocity and the whole machine is free to oscillate only in one mode, up and down for instance, the centrifugal force will cause the whole thing to be displaced from its rest position in proportion to the cosine of the angle formed between the arm of the rotor and the direction of displacement. As a cosine is merely a sine with a phase difference of 90°, the vibration resulting in this case will be capable of producing a 'pure tone' because the graph of the pressure fluctuations of a pure tone is a sine wave. In many machines, of course, the out-of-balance forces are much more complicated than those produced by a single rotor. One of the reasons is that machines in practice do not vibrate simply up and down, but in altogether six different ways: up and down, side to side, forwards and backwards, rotationally about a vertical axis (yawing), and rotationally about either horizontal axis (pitching and rolling). Consequently, in a machine with many rotating unbalanced parts, the overall vibration can be very complex.

There is an odd man out in the list of exciting forces which almost warrants a category of its own. This is vibration which is caused electromagnetically. It occurs most often in electric motors where the forces between the armature and the field magnet or the rotor and the stator cause distortion or vibration of the motor parts. Usually, though, the most important noise factor in an electric motor is aerodynamic, caused by the interaction of slots in the rotor and stator.

In transformers, magnetostriction occurs causing deformation of the core, and thus vibration with a fundamental frequency equal to the frequency of the A.C. current. There are usually many harmonics, of which the second is most prominent. In the case of 50 Hz current, the noise is therefore concentrated at 100 Hz for reasons of resonance in the oil and the casing, and the radiation efficiency of the unit as a whole.

Now we turn to the subject of impacts and percussion. Impacts, although efficient initial disturbances, are not alone very noisy. If the cymbal had been plastic and not brass, the noise of the impact between the stick and the cymbal would not have been very much. If you hit a stone wall with a clenched fist you will not make much noise. If you do the same on the side of a metal filing cabinet you will make much more of a din, but only because the other two factors of amplification and radiation are much more efficient.

Impacts occur in many machines. The largest ones are of course in presses, punches and stamping machines. Pneumatic drills make repeated impacts, by way of the bit, on to the road or whatever they are working on. The teeth or cutters of saws, routers and milling machines make high-speed impacts on the workpieces. In complex machines there are many percussive noises, the tapping of valve rockers on valves, gear teeth on one another, sprockets on chain links and many others.

Many of these initial disturbances, though, would make little noise on their own. Our recorder was not much good without its resonant body; a vibrating string makes hardly any sound without a sounding board; a plastic cymbal would be useless. Why is this? There are two reasons. Firstly, the initial disturbance in the recorder was of the random kind. By its very nature, random noise is not very efficient. Our vortices were all jostling about without any coherent reference to one another, with the result that a pressure increase from one was likely to be cancelled out by the pressure drop caused by the collapse of another. With the string, we have already seen how it cannot radiate much noise on its own because the air just slips round to the other side rather than be compressed. With the plastic cymbal, the bending of the cymbal by the impact is damped out too quickly for much of a noise to be made.

The amplification and radiation stages are essential. In the recorder, the resonance of the body had a feedback effect on the vortex formation causing the vortices to be formed in a more regular pattern. The fact that they were each formed exactly at the same rate as the column of air in the tube vibrated enabled them to work together constructively and not against one another in a random manner. The resonance of the string is of course just as important as the resonance of the recorder body. If you draw a violin bow across the edge of a table you will not have much joy, but even more important is the radiation efficiency, because there is the problem of the air slipping round the sides of the string.

However, the vibration of the string exerts a fluctuating force on the bridge of the violin, and this is transmitted into the wooden body. The size of the body compared with that of the string means that the cancellation occurs to a much smaller extent. It is dependent on frequency of course; the larger the surface, the lower the frequency which it can radiate efficiently, and the greater the intensity of the sounds which it does radiate. Hi-fi enthusiasts will know that a large speaker is necessary to reproduce low-frequency sound. By boxing in the back of the speaker the cancelling effect can be reduced still further, and this is done in the case of the bass-reflex cabinet.

As every point on a surface can be regarded as a separate noise source, a surface twice the size of another should radiate noise of twice the intensity, that is, 3 decibels more. This is true if the amplitude of vibration over the whole of the larger surface is as great as that of the smaller surface. Therefore, the noise of the chisel on the grinding wheel relies upon the radiation efficiency of the wheel and the chisel. If either have resonances of appreciable magnitude, which they undoubtedly will, the initial disturbance created at the point of contact between the chisel and the wheel will excite these resonances and enable sound energy to be radiated fairly well. There will also be a bit of aerodynamic noise caused by the air in the rough crevices in the surface of the grinding wheel being disturbed by the end of the chisel. It follows that the larger the wheel, the greater the intensity of noise and the lower frequency content. Not only will a larger wheel have lower resonant frequencies, but it will also be more capable of radiating low-frequency noise. The chisel, on the other hand, will have high-frequency resonances and because of its small size will only be a good radiator of high-frequency noise.

Resonance, though very good at amplification of sound, is not essential for a surface to improve the radiation efficiency. Remember that the violin body should have been designed by the craftsman who built it to have a resonant frequency below that of the lowest string. Surfaces which have no resonances, or are 'dead', can still be forced into vibration by the initial disturbance, and by virtue of their large surface area can increase the noise.

Where resonances do exist, their importance is very much dependent on how close a resonant frequency is to the forcing frequency of the initial disturbance. Where the two coincide, massive amplification can occur.

With initial disturbances that are the result of out-of-balance forces, quite often the amplification and radiating mechanism is the floor to which the machine is fixed. This is obviously a big surface which will be a good radiator of sound, particularly at low frequency. Other than this, of course, most machines have metal cases which can have almost as many resonances as a cymbal, and although they will not radiate much noise at the fundamental frequency, because often it will be too low, they will make up for it at higher resonant frequencies.

With impact sources, the amplification and radiation mechanisms are all-important, as we have seen. Without resonances in the object which suffers the impacts, the noise is merely the result of the sudden expulsion of air between the object and the striker, and only about two or three oscillations of the object lasting a millisecond or two. A resonant object will be excited by the impact and oscillate for anything up to a second or more, depending on the amount of internal friction or damping present which dissipates the sound energy.

In a press used for metal stamping, there are resonances in the metal of the press, the workpiece, and often the floor. In a circular saw, the impacts between the teeth and the workpiece, as well as creating aerodynamic noise from rotor/stator interaction, excite the blade of the saw, which is very like a cymbal, into sustained resonance. In fact, whenever any impact makes a reasonable noise, there is resonance somewhere. If you slap a wall with the flat of your hand, the airspace left in the slight hollow between your palm and the wall will behave like a cavity resonator. Cavity resonators are described in chapter 8 and are first cousins to resonant tubes.

Let us now turn to some important noise sources and see in detail how all these principles apply. Why are engines noisy? Why are diesels noisier than petrol engines? Why do aircraft scream on landing and roar on take-off? What happens in these machines to make them so noisy?

One of the most ubiquitous causes of noise in urban, and even rural, areas is the internal combustion engine. Little did Henry Ford know that his vehicle was to be a forerunner of one of the major nuisances of today. Of course, without the internal combustion engine the economies of the industrial nations would perhaps still be in their infancy, but, with the rapid growth of urban motorways and the escalation of engine sizes, noise and exhaust have become some of the most serious problems of pollution in cities.

The private motor car is not so much the culprit as the diesel-powered commercial vehicle. When Dr Rudolf Diesel in the last decade of the nineteenth century developed the diesel cycle, he unwittingly invented the noisiest internal combustion engine yet encountered. The diesel engine has become so popular because it can extract more work out of each heat unit than any other engine. In spite of the higher initial cost, the added weight and the lower power output for the same swept volume compared with the petrol engine, the diesel engine has proved to be more economical in the long run, and it is certainly here to stay in widespread use for years to come. The day when it will be supplanted by the gas turbine or the electric or nuclear engine is still a long way off. Let us therefore take a look at the causes of internal combustion engine noise and, in particular, diesel noise.

An internal combustion engine can be looked at as a means of converting noise into mechanical power – a very odd fact, but none the less true. It is not therefore surprising that, as no machine is 100 per cent efficient, our noise engine should release some of its noise energy to be heard by all around it. All the work in a reciprocating piston engine is done in the combustion chambers. The expansion of gases exerts a force on the faces of the pistons which is converted via the connecting rods and the crankshaft into a rotating source of power. If you measure the pressure in the combustion chamber of a typical diesel engine and plot the results against time, you will get a graph like figure 23 for an engine on full load at a fixed speed of 2,000 r.p.m. Now this graph has a very important property: the pressure function is periodic, and again

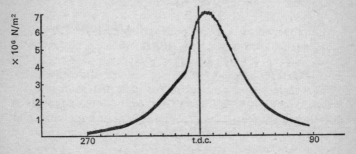

23. An oscillogram of diesel cylinder pressure

this means it can be subjected to Fourier analysis. The pressure graph can be broken down into a series of harmonics, the same as the harmonics in the musical instruments of chapter 3, but with more complex relationships between them. If the pressure graph is subjected to Fourier analysis, the harmonic components will turn out to be as shown in figure 24. This figure does not show the phase relationships between the components, but we are not interested in these in this case.

Now you can see why the engine is a noise-engine as much as a heat engine. What is going on in the combustion chambers is the generation of noise, of such a kind that the pressure pulses via the

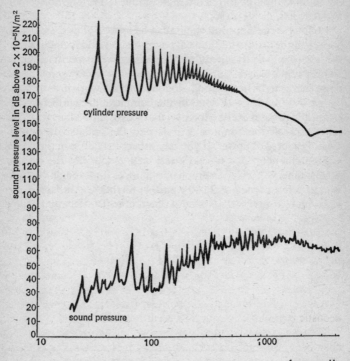

24. Cylinder pressure and narrow-band noise spectra

pistons are able to turn the crankshaft. As you might expect, the sound pressure level in the combustion chambers is immense, and is in fact over 220 decibels re 2×10^{-5} Newtons/m². If all this were to be released as noise outside the engine, there would be no power left, but fortunately the greater part is retained and put to good use. The pressure fluctuations on the outside surface of the engine are only about 0·001 per cent of the magnitude of those inside the combustion chamber, and at 1 m away the sound pressure drops a further 10 per cent, because of radiation losses.

Why does all the noise energy not escape? Why does any noise energy escape? Both these questions can be answered together. If we ignore for the moment the exhaust and inlet to the engine, the noise in the combustion chamber cannot reach the outside world without first entering the metal of the cylinder block and crank-case, usually cast iron. Not only does the noise have to enter the metal, it also has to re-enter air outside. We shall see in chapter 9 that in acoustics, as in many other branches of physics, one of the basic essentials for an efficient transfer of energy from one medium to another is that the resistances of the two should not be too different. The acoustic resistance, more commonly called the acoustic impedance, of cast iron is some 60,000 times as great as that of air. Consequently little sound energy is admitted to cast iron from air and vice-versa; the greater part of the sound remains inside the combustion chambers and much of the sound that gets into the cast iron never gets into the air outside.

There is, unfortunately, one means by which the cast-iron block can partly overcome this difficulty. If you take a cylinder block and crankcase and vibrate it in a laboratory, it will show that it has a multitude of resonant frequencies. Bending waves set up in the metal are reflected from all the changes in shape and thickness, causing standing waves at different points and frequencies all over the structure. There are of course harmonics of all these as well. The effect of resonance in a structure is greatly to reduce its acoustic impedance to sound of that frequency.

The result is that where a Fourier component in the combustion chamber noise coincides with a resonant frequency in the cylinder block, much more sound escapes. Looking again at figure 24, it is clear that this is happening with this particular engine at the frequencies of the third and seventh harmonics. The first, second, fourth and all the other harmonics are obviously not quite so 'in tune' with resonant frequencies in the cylinder block, and do not come through so easily. This engine of course is running at a fixed speed of 2,000 r.p.m., and with automotive engines of variable speed the pattern will change with every change of engine speed.

The cylinder block and crankcase are by no means the only radiators of sound. Vibration is transmitted into parts such as the timing cover and the valve cover, and these are usually even more resonant than the cast-iron parts. Large forces are also transmitted via the pistons into the crankshaft and so the crankshaft pulley becomes a noise radiator, as does the flywheel, although the latter is normally enclosed by the clutch housing.

We can now discover why a diesel engine sounds so different from a petrol engine, and is so much noisier. In the first instance, when a petrol engine has little or no load, the intake is throttled and the pressure rise in the combustion chambers falls dramatically, greatly reducing the amplitude of the Fourier components. In the diesel engine, on little or no load, there is no throttling of the intake at all, merely a reduction in the amount of fuel injected into the combustion chamber, with the result that the cylinder pressure barely drops.

However, more important is the fact that in a petrol engine the time it takes for combustion to spread from the spark plug to the edges of the chamber gives the pressure curve a fairly smooth profile. Not so with the diesel. When the metered squirt of fuel is injected into the combustion chamber, pure air has been aspirated and compressed to such an extent that its temperature has risen well above the ignition temperature of the fuel oil. The fuel is all burnt virtually simultaneously, and so the pressure graph shows a sudden sharp rise at this point. When it comes to a Fourier analysis of a diesel pressure curve and a petrol one, the diesel is found to contain higher harmonics of a greater intensity because of the different shape of the curve, and this accounts for the characteristic knocking sound of the diesel. The other important result of the difference in the Fourier series is that the rate of increase in noise with increase in speed for the petrol engine is higher than for the diesel engine. For the petrol engine a tenfold increase in speed gives a

50 dB increase in noise, whereas the same increase in speed for a diesel engine only raises the noise level by 30 dB. If you could rev a petrol engine and a comparable diesel engine hard enough, a speed would be reached where the noise level was the same from each of them.

On the other hand a tenfold increase in engine volume raises the noise level by only 17 dB. This means that if you use a slower engine and increase the cubic capacity you will have less noise. A fast-revving small engine will always be basically noisier than a slower-revving large engine developing the same power. When you also take into account the fact that the slower-running engine produces lower-frequency noise, which falls in the part of the audible frequency range to which the ear is less sensitive, the result is altogether quieter.

There are several other sources of noise on an engine, some independent of, some linked with, the combustion of the fuel. Every time an exhaust valve is opened, there is a sudden release of pressure into the manifold. The pulsing of gas into the manifold during every revolution of the engine can be broken down into a series of harmonics just like the graph of the pressure inside the combustion chambers. The manifold and exhaust pipe will have their own resonances and harmonics, and where these coincide with the frequencies of the exhaust noise, amplification occurs. Fortunately most engines are fitted with fairly efficient exhaust silencers, the design of which is described later. Even if an engine does have a noisy exhaust, it is usually only because of economics that a poor silencer has been used, and it is always technically possible to reduce exhaust noise to below the level of the engine noise itself. The latter is by far the more difficult problem to overcome.

Inlet noise has to be reckoned with also, and sometimes is overlooked by engine manufacturers. When the inlet valve opens, there is still a little pressure in the cylinder which causes a small pulse of gas to come back into the inlet manifold. Almost immediately the reverse happens and gas is sucked into the cylinder. Noise from the inlet is often predominated by resonance of the manifold, sparked off by the gas oscillations in the inlet ports.

In petrol engines, both exhaust and inlet noise is much reduced when the engine is not under load, but with the diesel there is not much difference. In a petrol engine, exhaust noise goes up when the engine is used as a brake, for instance when the car is running downhill and is turning the engine with the throttle closed. Fuel injectors play a supporting role in diesel engines and some petrol engines. Their noise is very often less than that of the engine. Timing chains, valve gear and the transmission all contribute extra components to the noise.

There is another instrument in the under-bonnet orchestra which plays a much louder part than many people realize, the engine cooling fan. Fans are inherent noise makers anyway, and merit their own position in the list of machines described in this chapter. In motor-car engines, they produce a noise which at some engine speeds is as loud as the engine noise itself. Fan noise rises by 55–60 dB for a tenfold increase in engine speed, and so, as an engine revs up, engine noise loses ground to fan noise, which also takes to absorbing power in, in most cases, overcooling the engine. Removal of a fixed fan and substitution of a thermostatically controlled unit either electrically driven or fitted with a magnetic clutch not only cuts down engine noise but will give you back a few brake horse power at high engine speeds. Unfortunately the saving in fuel is small compared with the cost of these units, but the extra mile or two per hour or gallon and the less noise make them welcome accessories.

Although fans by no means rank high in the charts of principal noise makers, it would be a good thing to look at them now, because when we come to that acoustic ogre the jet engine we shall find that they have a lot in common. They are also major noise sources in the world of ventilation and air-conditioning.

A fan is, of course, a machine for moving air from one place to another. It will normally be one of two types: axial and centrifugal. Figure 25 shows typical fans of both types. An axial fan has much in common with an aircraft propeller, and in fact axial fans working against little or no static pressure that only require simple housings are generally known as propeller fans. A centrifugal fan spins the air and centrifugal force causes a radial flow of air which is funnelled by a scroll into a discharge at right angles to the shaft of the fan. These fans may have forward-curved, backward-curved or

25. An axial fan (top) and a centrifugal fan (below)

radial blades, giving them different performance characteristics. In any fan, the basic noise-making mechanisms are the same, producing two types of noise: vortex noise and rotation noise. We came across both of these earlier in the chapter.

The vortex formation is a source of broad-band noise with no discrete frequency components and forms the 'accompaniment' to that other type of fan noise: rotation noise. If you select a point in the air close to the path of the rotating blades, each time a blade goes by the air at that point will be given an extra shove, causing a sudden pulse of pressure. In most fans the blade-pass note is within the audible range, and if you plot a graph of the pressure pulses at any point you will get another periodic waveform. Once again, because it is periodic, the wave is susceptible to Fourier analysis and can be broken down into a series of harmonic components. The fundamental frequency will be equivalent to the blade-pass frequency of the fan. There will then be a multitude of higher harmonics.

This rotation noise is again made even worse when the blades pass close to stationary objects such as vanes or motor supports in an axial fan, and the cut-off in a centrifugal fan: the air is suddenly 'clipped' as the blade passes by, causing even more rotation noise. This effect is a function of the clearance between the rotating and stationary part, and happens in all rotating machines as well as fans, particularly turbines and electric motors. In the case of vanes in an axial fan, if there are the same number as there are blades, the blade-pass note will be amplified; if the numbers are different, the overall increase in noise will be less. Still more noise can be added to the vortex and rotation noise if, as often happens, any part of the fan, fan supports or casing is allowed to resonate. One of the worst cases occurs when the blades themselves are resonant, and if the misfortune occurs that a resonant frequency or harmonic in a blade or vane coincides with the blade-pass frequency or a harmonic, the two will of course work together and the result will be another increase in noise. Some propeller fans have large, nearly flat, sheet-steel stampings for blades, and of course these are the most resonant of all. Many axial fans now incorporate blades of thermoplastic or similar materials which have high inherent damping and thus are very poor resonators. We shall be taking a look at the properties of different materials in chapter 12.

Fan noise can be fairly accurately worked out from theory. The overall sound power level is a function of the capacity, static pressure and absorbed horsepower. Knowledge of any two of these factors is enough. If we know the sound power-level of a fan, we can get an idea of the frequency distribution of the sound for both axial and centrifugal fans. Centrifugal fans nearly always have the bulk of their noise at low frequency in the 63 Hz octave band, and the rest of it spread progressively less in higher bands. The average is a 5 dB drop per octave increase in frequency, but with some fans, in particular those with backward-curved blades, the greatest amount of sound energy is at present at the blade-pass frequency.

With axial fans, rotation noise is prominent, but the proliferation of harmonics goes to make a fairly evenly distributed sound spectrum, falling off a few decibels at the low- and high-frequency ends. They are generally easier to silence than centrifugal fans because they have so much less low-frequency noise, which can be the very devil to reduce. On the other hand, when fans are incorporated in a duct system, inherent losses in the ducting usually do a good job on the low frequencies in a centrifugal fan, and do little to reduce axial fan noise. The latter sounds much worse because it is full of pure tone components, whereas the residue of the centrifugal fan is all vortex noise and much less offensive.

Now for the most infamous noise polluter, the jet engine! The noise of a jet flying overhead is among the loudest noises the man in the street (or in bed) is likely to hear. It perhaps causes more distress to more people than any other, and is gaining on its nearest competitor – road traffic. So great is the problem that the jet engine now enjoys the almost unique position of having noise as one of the first priorities on the designer's list. However cheap, efficient, light or economical, no jet engine will see service in any civil aeroplane if it makes too much noise.

'All jet engines make too much noise!' you are probably saying. Everything is relative, and it would be unfair not to credit some engine designers with no small achievement in the reduction of jet noise. Before going on, let us also remember that first cousins to jets, gas turbines, are in widespread use as stationary and marine engines, notably in some of the most modern power stations. They are also just beginning to appear as automotive engines. Just because the public has been protected from the noise of these engines by the skill of acoustical engineers who do not have to worry about whether their silencers can fly, they must not be forgotten.

Newton's Second Law states that 'To every action there is an equal and opposite reaction'. Both rocket motors and jet engines rely entirely on this principle. The rocket motor carries with it liquid propellant which is burned in a combustion chamber. A

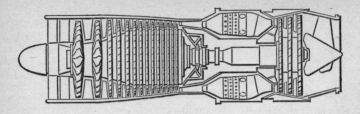

26. A turbo-jet

great amount of heat is released in the burning of the fuel, and also large volumes of gas. The newly formed gas expands at a great rate and then leaves the combustion chamber by the only means open to it: the nozzle at the rear. The result is that a large mass of gas is constantly accelerated from the rear of the rocket, and can be likened to an unbroken stream of bullets from a gun. Anyone who has fired a gun will be painfully familiar with the process of action and reaction in the recoil or kick that occurs after firing every round. If the gun fired a continuous bullet, the kick would become a continuous force against the shoulder of the user. This is purely the reaction against the action of expelling the bullet. In just the same way there is a reaction to the expelling of the mass of exhaust gas from a rocket motor, and it is this which propels the rocket.

A jet engine is really a rocket motor which does not carry all its gas with it, but makes use of the abundant gas all round it: air. The simple turbo-jet engine has, like a rocket motor, combustion chambers and an exhaust nozzle through which heated gases are accelerated to provide the thrust. The heated gas is produced in the same way as that in the combustion chamber of a piston engine: air is compressed, vaporized fuel is added and the mixture is ignited.

In a turbo-jet, though, this is a continuous process; the compressor is an axial, multi-bladed, multi-stage device like several highly sophisticated axial fans in series; the fuel is continuously injected into the combustion chamber along with the compressed air, and once the engine is started ignition is spontaneous and continuous. Something has to drive the compressor, so a turbine is fitted downstream of the combustion chamber to borrow some of

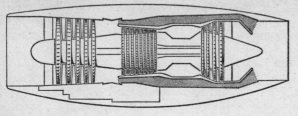

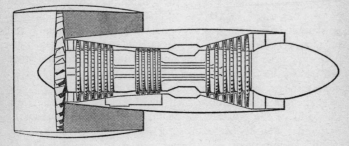

27. Turbo-fans

the power in the expanding gases and drive the compressor. A turbine is like a fan in reverse, or a sophisticated windmill, and it can drive the compressor direct by being mounted on the same shaft.

The noise of a turbo-jet has several sources. The most powerful noise source is the turbulent mixing of a high-speed gas stream with the atmospheric air around it. Our old friends the vortices are back again, and we are now familiar enough with them to see fairly easily how a gas stream can be noisy. The interesting thing is that if you set up a test rig and produce a stream of cold air which in all other respects is identical with the discharge from a turbo-jet, the noise it produces is almost exactly the same. The sound source is not the discharge orifice, but extends some distance away in what is known as the 'mixing zone', where the mixing of the high-velocity turbulent air with the atmospheric air takes place. You might think that a stationary jet would be noisier than one moving at speed because of the greater gas-stream velocity relative to the surrounding air. To a certain extent this is true, but what actually happens is very much more complicated and can result in an increase in high-frequency noise.

The noise from the jet efflux does not radiate with the same intensity in all directions. By far the greater part is concentrated to the rear, and the directivity pattern is heart-shaped; the highest noise level occurs to the rear about 40° off the axis of the engine.

Next in importance to jet noise in the turbo-jet comes compressor noise. The first stage in the conventional engine is a set of guide vanes, after which comes the first rotating stage, followed by an alternation between rotors and stators. There may be some fifteen of each. The interaction of rotors and stators creates high-frequency siren-like noise, containing harmonics as usual, which radiates to the front of the engine. This is the scream which an aircraft makes when coming in to land. Compressor blades made of conventional metals have their own resonant frequencies, and thus add to the noise. Once again, when blade resonance or harmonics and the rate of rotor/stator interaction coincide, amplification occurs.

The turbine also screams, but this noise radiates rearwards and tends to be overshadowed by the jet noise.

The other radiator of noise from a jet engine is the casing of the engine itself. Noise from inside, from combustion and from the compressor and turbine both penetrates the casing and sets the casing into vibration, thus causing it to radiate noise. Compared with noise from the jet efflux, casing noise is very small indeed.

Great strides have been and are being made in the design of jet engines, and the basic turbo-jet engine is becoming a thing of the past. Several of the noise-generating features of these engines can be improved; the gas velocity is reduced and the mixing improved in the modern turbo-fan engine. In these engines the first stage of the compressor is enlarged, and the engine spills much of the air around the casing.

However, recently the public have come to discover that the engines on aircraft are by no means the only cause of noise pollution. In fact, a totally 'silent' aircraft could create a major noise problem, if it were to fly faster than sound. Sonic boom is a new type of noise which could dwarf all others as a major cause of national dissent. What, then, is sonic boom? Early films made at about the time military aircraft first flew at supersonic speeds dramatized the phenomenon of 'breaking the sound barrier'. Begoggled pilots jutted their jaws and gritted their teeth as the screen shuddered and beads of sweat ran down their foreheads when, like Armageddon, the consummate, nerve-rending moment came and they crashed through the sound barrier. I am told that if you are not on the ball you can in fact reach Mach 1 without noticing a thing.

It is the people on the ground who suffer. When a solid object is in relative motion with air, there is a rise in the pressure of the air immediately in front of it. At subsonic speeds, this pressure is constantly being dissipated because it travels away as a wave front at the speed of sound (figure 28). However, if the relative velocity between the object and the surrounding air is greater than the speed of sound, the pressure cannot get away ahead of the object. What happens is that the pressure wave piles up on itself (figure 28) and forms a cone-shaped shock wave. The cone spreads out and consequently the shock wave eventually reaches the ground, being heard as a bang or boom. Contrary to many people's ideas, sonic boom occurs in the wake of an aircraft *all the time* it is flying faster than sound, not merely as it 'breaks the sound barrier'. Because temperature lessens with increasing altitude up to the troposphere, the temperature gradient (chapter 7) causes the path of the shock wave to be bent upwards so that it does not always reach the ground. The Mach number of an aircraft travelling at 10,000m has to exceed 1·3 before the shock wave has significant magnitude on the ground. Its actual waveform is N-shaped (figure 29) and although in reality a shock wave is created whenever there is a change of shape on the aircraft, such as at the point where the wings join the body and at engine cowls, the tail plane and fin, these intermediary shocks coalesce with increasing distance so that by the time they reach the ground there are only two major pressure jumps, caused by the nose and the tail giving rise to the N-shaped wave. This is why when you hear a sonic boom it is usually two bangs in quick succession.

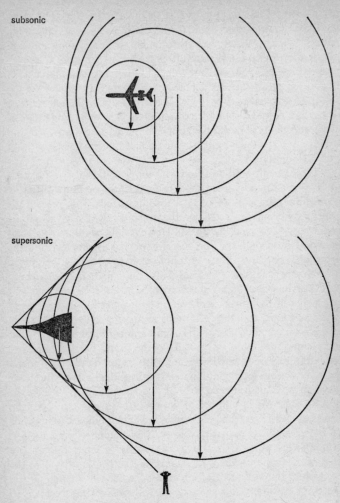

subsonic

supersonic

28. Air pressure waves at subsonic and supersonic speeds

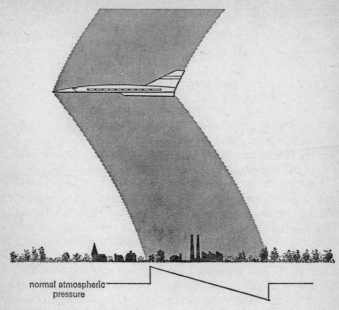

normal atmospheric pressure

29. Sonic boom

With increasing altitude the intensity of boom tends to diminish. One must also remember that at the altitudes at which supersonic aircraft fly the air is somewhat rarefied, which results in less of a shock wave. One of the reasons why supersonic aircraft have needle-sharp noses is that this reduces the pressure wave in front of the aircraft.

7 From A to B – sound in the open

What happens to the sound that we now know so much about after it leaves the source? We have looked at several sources, and know all about ears, but what happens in between? The answer is rather more interesting than you may think.

Let us go back for a moment to our early acquaintance, the pulsating balloon, and imagine that it is suspended in mid-air. In the last chapter a remark was made which will now turn out to be rather more significant than it seemed at the time. Every point on a vibrating surface can be regarded as a separate source of sound. If we go back still further, to chapter 2, we can unearth the description of how a sound wave is transmitted by molecules or particles compressing the 'springs' between each other, and when the springs build up potential energy they impart it to the molecules in front, which start to move and gain kinetic energy. But we overlooked one thing. Pressure in a fluid never exists in one direction only. If you exert a downward pressure on the piston in a tyre pump and make a hole in the side, air will hiss out just as fast as if you make a hole in the bottom. When you think of a typical sound wave of about 200mm length, and draw a graph of the pressure along the direction of travel of the wave, taking for example a sine wave, you can see that the section of air under compression at any one moment extends for half the wave length, 100mm, not just the distance between the molecules. The air molecules are by no means lined up in a regimented pattern so that the springs between them are all acting in a straight line, but are rather in random positions with the intermolecular forces acting in all directions. The result is that if you could freeze the air with a sound wave in it and insert a directional pressure gauge in the middle of an area of compression, you could turn it in any direction and get the same reading.

What does this lead to? Every point on the surface of a noise radiator can be regarded as a separate noise source, and therefore exerts pressure fluctuations in every direction. Figure 30 shows a

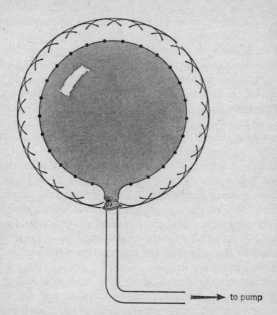

30. A pulsating balloon showing Huygens' method of wave construction

few selected points on the pulsating balloon as it is expanding. This virtually means that each point starts off a hemispherical sound wave of its own. What is more, because the pressure in each wave acts in all directions, every point on this wave front can be regarded as a new source of sound. This is known as Huygens' method of wave construction.

You could be forgiven for thinking that the result of all these myriads of tiny theoretical waves would be a complete boggle of sound waves. In fact, if you look at figure 30 you will see that it all works out very nicely. Remember that, if you have two or more sound waves at any point, their pressures or intensities are added together, a process known as superposition. Of course, if these quantities are expressed in decibels, you obviously have to go through the normal procedure for addition of sounds. If the little hemispherical waves from the points on the surface of the pulsating balloon are all added together, the result is a new wave front which is spherical and concentric with the balloon. Furthermore, every point on this new wave front is a new source of sound, but when the little secondary waves from these points are added together, another concentric sphere is the result. In fact, there is no need to bring Huygens into it at all – it would be just as valid to say that the spherical balloon radiated ever increasing spherical waves of sound travelling at 334m/sec. However, getting a true understanding of wave construction is the only way in which one can understand many of the things which sound waves do.

Once these spherical waves are sent out by the balloon, they expand in size as they travel away. The result is that the sound intensity gets very much diluted. Every time the distance from the source is doubled the area of the spherical wave increases four times, causing a drop in sound intensity, and thus sound pressure, of 6 dB. In other words, the sound intensity is inversely proportional to the square of the distance from the source. This is called the inverse-square law.

If one knows the sound power level, it is very simple to calculate the sound pressure level at a distance if the source is radiating sound spherically with the same intensity in all directions. We have seen that to all intents and purposes sound intensity level is equal in decibels to sound pressure level, and the intensity at a distance r from the centre of the source is simply reduced by

$$\frac{1}{4\pi r^2}$$

To turn this into a decibel change, the formula becomes:

$$\text{Decrease in sound level} = 10 \log_{10} \frac{1}{4\pi r^2} \text{ dB}$$

However, life is not always so simple, because not many sound sources are so convenient in radiating sound completely spherically. Let us get rid of the balloon once and for all and take a more sophisticated sound radiator, a vibrating steel plate. This is where the complicated method of wave construction becomes useful. Look at figure 31. The little spherical waves from the individual

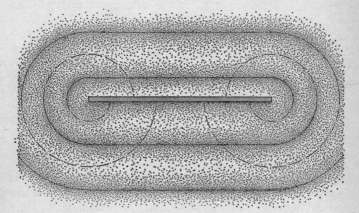

31. Sound waves from a vibrating steel plate

points cancel each other out at the edges of the plate because the points on one side of the plate are exactly 180° out of phase with those on the other side. In the middle of the plate the 'envelope' of the wavelets is not a sphere but a flat surface or plane wave. The result is that almost all the sound intensity is concentrated in the direction at right angles to the plate, and if you use the basic inverse-square law you will come unstuck.

As the waves travel farther away from the plate, or if the wavelength gets longer, the cancellation of the waves at the edge becomes less complete. The result is that a few metres away, in what is known as the far-field, the radiation changes to a near spherical pattern, and the inverse-square law becomes valid again, provided we know the proportion of the sound intensity which is radiated in the direction of interest. This is known as the directivity factor (Q_θ).

This directivity factor can be calculated for any angle 'θ' for something as simple as a plate, but it is much easier in most cases to measure it. Then if we know the sound power level and the directivity factor, we can simply calculate the sound pressure level at a certain point 'r' metres from the centre of the source, and 'θ' degrees off the axis from the formula

$$SPL = SWL + 10 \log_{10} \frac{Q_\theta}{4\pi r^2}$$

Working out the effect of distance on a sound level does not seem too bad, particularly if the two points are in the same direction so that knowledge of the directivity of the source is not necessary. This is only so out of doors with no near-by objects, no wind and equal temperature at all distances from the ground. Unfortunately this condition never exists, and there are several flies in the ointment. First of all, there is nearly always a near-by surface to reflect sound, even if it is only the ground. If the reflected path is not much longer than the direct path, the two sound waves, reflected and direct, can be nearly equal and the intensity or pressure can thus be doubled giving a 3 or 6 dB increase. If there are multiple reflections the sound level can go up even more.

However, sometimes more important is the effect of wind and temperature variations with altitude. The atmosphere is never completely still and of equal temperature at all altitudes. Many people will have noticed that it is difficult to hear upwind, quite apart from the fact that wind noise itself is present to mask the sound one is listening to. Some people put this down to the fact that the wind is 'blowing' the sound back, but this is not quite true unless the wind reaches the speed of sound, in which case you would not be standing there.

Wind has several effects on sound, the first being a simple matter of velocity vectors. Figure 32 shows the case of a sound wave travelling upwind at a slight angle, and the parallelogram law is used to work the resultant direction and velocity. The other effect that wind has on sound propagation is due to the fact that wind speed near to the ground tends to be lower than that higher up. This means that the sound travelling upwind will have a greater overall velocity close to the ground.

Before explaining the effect that this has, let us bring in the question of temperature. During the daytime, the sun radiates heat

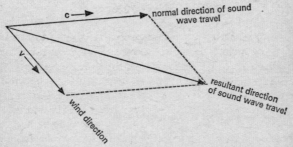

c ⟶

normal direction of sound wave travel

v

wind direction

resultant direction of sound wave travel

32. The effect of wind on the direction a sound wave travels

which warms the ground. The ground then warms the air above it, and the process of convection then warms layers of air successively higher from the ground, but to a lesser extent. On a calm, sunny day this negative temperature gradient, known as a temperature lapse, can be appreciable. At night, though, the ground gives up its heat to the air and no longer receives heat from the sun. Convection carries the air warmed by the ground to higher altitudes and on a clear, calm night there can be a large positive temperature gradient or temperature inversion.

The velocity of sound is faster in warm air than in cold air. When there is a temperature lapse, the velocity of sound is slower the farther the distance from the ground. Exactly the same occurs when there is a positive wind gradient and the sound is travelling upwind. The wind velocity has to be subtracted from the sound velocity to give an overall sound velocity that decreases with altitude.

How does this affect sound propagation? Look at figure 33. This shows a wave front travelling in a positive wind gradient and a temperature lapse. The top of the wave front is passing through cooler air or is making less headway against a faster wind than the bottom of the front. This has the effect of tilting the wave front

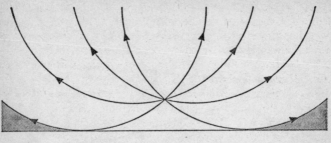

temperature decreasing with height

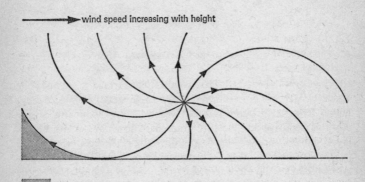

wind speed increasing with height

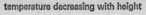
sound shadow   → direction of sound propagation

33. The effect on sound propagation of a temperature lapse and a positive wind gradient

upwards. If you are rowing a boat and pull harder on one oar than the other, the bow of the boat will turn towards the other side.

Now look at figure 33 and see what the overall effect of this is. Upwind of the source, or on either side of it in a temperature lapse, the curved path of the wave fronts mean that the ground has a shielding effect, causing an acoustic shadow beyond. The shielding is not complete because of diffraction of sound into the shadow, a phenomenon we shall be looking at soon. However, the result is that beyond the critical distance between the source and the point when the lowest sound wave grazes the ground there is a big reduction in sound intensity. This is much more noticeable at high frequency, again for reasons of diffraction. Downwind, on the other hand, or in a temperature inversion, the wave fronts are bent the other way, giving completely the opposite effect.

There are so many variables that a theoretical estimate of the effect of gradients would be unreliable, but to give an idea of the order of things, the reduction for a receiving point one kilometre upwind of the source with a windspeed of some 20 k.p.h. would be around 30 dB at 1 KHz. Wind, of course, is often turbulent, and the effect of this is to scatter the sound waves and deflect them into the shadow zone. The turbulence also tends to reduce the intensity of the wave.

Gradients are not the only things which affect sound propagation out of doors. An important factor with long distances and high frequencies is the effect of air viscosity. As the air particles oscillate to and fro, there is a certain amount of friction between them and their neighbours. Friction always dissipates energy, and at high frequency when the particle velocities are highest it can be significant. At a distance of one kilometre from the source, a 10 KHz sound could be attenuated by some 40 dB, in addition to other reduction, for instance due to distance and the inverse-square law. The ground can also absorb sound energy, a process which the next chapter looks at, and if the terrain is hilly or wooded, or covered with snow, this effect can be quite important.

So far, we have not allowed for our sound waves encountering much more than the effects of wind and temperature. As few of us live in a desert, most of the sound we hear out of doors meets many obstacles which are much more solid than the wind. When a sound wave strikes a surface, we will not know what happens to it until after the next two chapters, but, as most surfaces do not go on for ever, it is important that we should know what happens to the sound that gets past the edges of these objects.

Imagine a wall which lets no sound at all through it and which has a long straight edge to it. What happens to a sound wave which just gets past the edge? To many people the answer is that it will travel straight on, and form the edge of an acoustic shadow the other side. After all, that is what happens to light, and sound, we are told, travels in straight lines just like light. This is not true for either sound or light.

Remember Huygens' method of wave construction from the beginning of the chapter. Every point on a wave front can be regarded as a separate source of secondary waves. Every point on a wave front can be regarded as a separate source of sound sending out little spherical waves, and because of the continuous line of points all these little waves combine to form a replica of the original wave front, a short distance ahead. Now if a sound wave front meets the edge of a wall, part of it will go past the wall and part of it will not. The points near the end of the section of wave front which gets past the wall will have no companions on one side of them any longer. The convenient combination of little spherical waves which went to make up the main wave front as it moved forward can no longer occur on one side of the wave front. The result is that the truncated end of the front becomes a new source of sound radiating spherical waves at right angles to the path of the front, down into the sound shadow. They still combine together, but the result is very different indeed.

Look at figure 34(a). The progress of a series of truncated wave fronts past the wall has been charted. Also shown are the positions which the new spheres of sound, sent out by the points at the chopped-off end of each front, have reached at a particular moment. For simplicity, only a few points along the path of the front have been shown, and in reality there is a continuous line of them.

What has happened now? In the direction at right angles to the path of the main wave fronts, the secondary waves cancel each other out completely. Points of maximum compression of some coincide exactly with points of maximum rarefaction of others. On the other hand, the nearer you get to a position on the line down which the edge of the front is travelling, the less complete the cancellation of secondary waves becomes. The effect is that at point 'a', for instance, the shielding effect of the wall is not much use because the secondary waves combine fairly constructively and produce as much sound as the main waves. However, down at point 'b', the secondary waves, for pure reasons of geometry, combine destructively, and to a large extent cancel each other out.

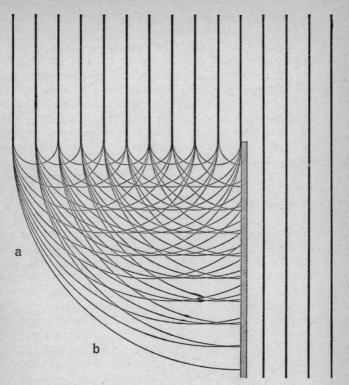

34. (a) The result of a series of wave fronts passing the edge of a wall

The result here is that the wall shields a listener at 'b' from most of the sound.

Now look at figure 34(b). This is the same as figure 34(a) but for a lower-frequency sound, which therefore has a longer wavelength. The difference in the two figures is quite remarkable. With the longer-wavelength sound the cancellation of secondary waves is not nearly so effective until the observer gets much deeper into the sound shadow. If we were to draw yet another diagram for a sound wave of very short length indeed, the cancellation of secondary waves would be effective almost up to the ends of the main waves, with the result that high-frequency sound behaves

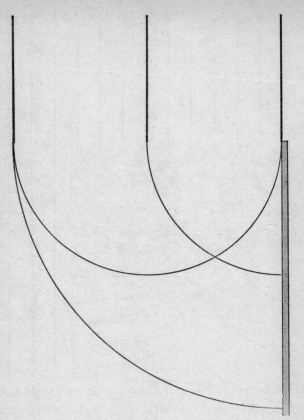

(b) The same situation as in (a), but with sound of lower frequency

much more like light than low-frequency sound. This process is called diffraction, and occurs with any type of wave, even light. With light, though, the wavelength is so short that the effect of diffraction is very difficult to detect.

As we have already seen, very few sounds contain only one single frequency. The effect of diffraction on a complex wave is exactly the same, but the low-frequency components are diffracted to a much greater extent. In order to discover how great the effect

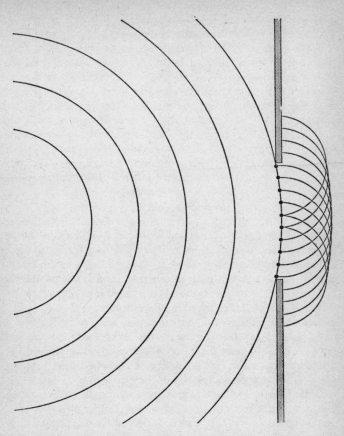

(c) Radiation of sound from an orifice, using Huygens' method of wave construction

of diffraction will be, the geometry of the situation must be worked out. The distance of the source and the receiver from the wall and the heights of all three are essential. Diffraction, of course, explains why you can hear things which are out of sight, and occurs in all sorts of situations and circumstances in acoustics. When we were on the subject of shadows due to wind and temperature gradients, it was mentioned that diffraction of sound into the shadow was a factor which reduced the overall effect.

This business of wave cancellation is important whenever there is an uneven distribution of sound. You may have noticed the similarities between the description of diffraction and the explanation of the directivity of a vibrating steel plate. When you are dealing with ventilation duct terminations, plane waves in the duct emerge into free air, and the same thing occurs as with the steel plate. The high-frequency sound tends to be beamed to the front, but the cancellation of the secondary waves at the sides of the plane waves becomes less complete at low frequency and thus low-pitched sound from the duct is much less directional. In the jargon of physics, the phenomenon that we have been calling 'wave-cancellation' or combination of waves is called 'interference'. Whenever two waves pass the same spot, interference occurs. This happens in a vast number of cases; we first saw it occur in the resonant tube, when the reflected wave and the initial wave caused interference and a standing wave pattern was the result. Special resonant tubes used in laboratories to measure the reflectivity of a substance at one end are called interferometers.

Sound is of course reflected in many more ways than by reaching the end of a tube, the most common being when it meets a large, flat, solid surface. Let us put the cart before the horse and have a look at some of the effects of reflection before going on to discuss the causes in chapter 8. Figure 6 showed what happens when a forward-travelling wave combines with a backward-travelling wave. The result of course is a standing wave which pulsates on the spot and does not move in either direction. This figure applied to a plane wave which was travelling along a tube. However, plane waves are in reality not all that common, and we are far more likely to meet that other type of wave, the spherical wave. What happens when one of these is reflected?

Figure 35 shows a graph of a spherical wave which has been frozen so that we can plot the points of maximum compression and maximum rarefaction and join them together. Now introduce a large reflective surface and add in the reflected spherical wave. Once again we have interference, and by superposition the compressions or rarefactions of one wave combine with or cancel

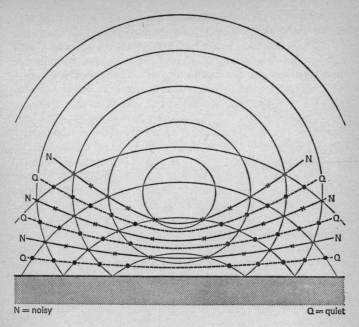

N = noisy   Q = quiet

35. Interference from a reflected wave

those of the other. The result is really rather amazing. There are
zones where the two waves always combine constructively and
increase the intensity of the wave, and zones where they always
combine destructively and reduce the intensity to zero. With low-
frequency sounds of long wavelength, these zones are quite large,
and that is the reason why you can walk about in a sound field
where there is low-frequency sound and reflective surfaces, and go
from areas of high intensity to areas of comparative quiet. On the
other hand, with high-frequency sound the zones are very small,
and if you are listening to a high-pitched sound in a reflective area
you can find that one ear is in a high-intensity zone and the other
in a quiet zone. The effect can be quite strange. If you move your
head from side to side you can reverse the effect.

Once again, the example that has been used is that of a pure tone.

If in reality there is a whole range of tones, the interference pattern is more elaborate. With random noise, where there are no periodic components, no neat pattern occurs and there are no zones of high and low intensity because the wavelength and amplitude are forever changing in a random way.

8   Sound absorption – holes in the ceiling

Walk into any modern building and look up; the ceiling will most probably be covered with thousands of little holes. It is a familiar sight, an acoustic ceiling, but what does it do? The sheets of holes seem almost blessed with magical properties to 'suck up' noise, or to spray 'quiet' about the room, and have caught the imagination of some who eagerly construct sound-proof walls from peg-board with subsequent sudden dismay.

They are of course nothing of the sort. An acoustic ceiling is no more than a poor reflector of sound: when sound travelling through air meets a hard, solid surface, it is reflected in much the same way as light from a mirror. Dark-coloured paper is a poor reflector of light; you cannot see yourself in it, and if you paper the wall with it you cut down the light. An acoustic ceiling is no more than acoustic brown paper.

What really happens when sound is, or is not, reflected? We tend to take this phenomenon for granted, but to understand the different ways in which sound absorbers work, and to get a better grasp of their usefulness or uselessness, it is well worth going back to the subject of reflection and taking a much closer look.

Many flat solid surfaces are virtually acoustic mirrors. A thick granite wall reflects 99 per cent of the sound that strikes it; other solid surfaces usually reflect about 95 per cent. Remembering about decibels and loudness, one can see that there is no noticeable or even easily measurable drop in the level of the reflected sound wave. What is in fact happening? A sound wave, we know, is merely a compression wave followed by a rarefaction which is passed through air at great speed, rather like the effect of a shunt on a train of loosely coupled railway wagons. There is no net movement of air, but in order for the compression to build up at each point along its line of travel, the air molecules crowd closer together, and have to move slightly to do so. As air is elastic they soon spring back and overshoot their original positions to find that they are spaced more widely apart, causing the rarefaction that follows. In practice, few sounds are an orderly alternation of compression and rarefaction and the molecules are jostled about more like a rush-hour tube traveller.

Imagine now what happens if the compression wave meets a wall of granite. As the compression waves are passed from molecule to molecule they meet a layer of air next to the solid surface of the granite. The wall is heavy, solid and stiff. When the molecules of air next to the wall get a shunt from their neighbours behind, instead of passing the shunt on to more air molecules in front they find that they are just pushed up against the wall and the compression builds up. Because air is elastic, they soon spring back with the extra force of the built-up pressure and so start a compression wave off in the opposite direction, almost as strong as the one that squashed them against the wall in the first place.

Granite is also elastic, although very stiff indeed, and does deform under pressure. It is of course the resistance to deformation, or strain energy, which when you stand on the granite so conveniently furnishes your feet with just the right amount of force to stop you sinking in, and will also cause another piece of granite to bounce, but that is another story. In the same way, there is a microscopically small deflection of the granite under the stress from the layer of air which is under such difficulties on the surface. This deflection or strain sends off a sound wave into the body of the granite, and that is where most of the missing 1 per cent of sound energy goes.

This missing 1 per cent has been absorbed, in that it has been removed from the incident wave, only 99 per cent of which is reflected. Part of it went into heat created when the granite was deformed, but most of it travelled away in the granite, and will reappear in the next chapter. If the wall is thinner, lighter and weaker, it will deform more under pressure, and when the wall is thin compared with the wavelength of the sound in the wall material, the whole wall gives under pressure from the incident sound wave. In these cases, which are the ones most frequently met with, it is not so much the stiffness of the wall material which counts; the governing factor is the inertia of the wall. The heavier the wall, the greater the inertia and the less it will move when subjected to a force. If you kick a football, it will shoot away from your foot; if you kick a rock, the rock will stay put and you will not be able to kick any more footballs for a day or two. You could say (among other things) that your foot was reflected from the rock and it was the force of the reflection that bruised your toes. The heavier the rock, the bigger is the bruise.

Except in special circumstances even relatively light non-porous walls reflect at least 90 per cent of the sound and this is much the same at all frequencies. Reflection of this order is quite clearly undesirable in many cases, and the story of sound-absorbing devices is all connected with persuading the reflecting surface to take in more of the sound and not to send it back.

If we look again at the plight of the boundary layer of air, an important point arises. If you drop a weight on a spring, it will bounce up again. If you fit a damper to the spring, like a car shock absorber or damper, a device through which friction or viscous forces resists movement of the spring up and down, the weight will hardly bounce at all. Most people will have noticed the way in which the front of a car bounces up and down when the dampers are worn out.

Air, as we have seen, can behave like a spring, and it is this property which causes it to reflect a compression wave or to carry waves at all. If the air can be damped like the spring the effect will be the same. As the compression builds up on the reflecting surface the air molecules are crammed closer together and there is a small movement of air particles towards the reflecting surface. If a little friction is introduced, to resist particle movement, what have we got? Damping! The air becomes less of a spring because in overcoming the friction resisting the movement of particles heat is generated and the energy needed to produce the heat is lost from the sound waves. In case you should now be devising a plan to heat your house by absorbing noise from aircraft, remember that even in a noise of 100 dB the energy flow amounts to only 0·01 watts per square metre.

Creating the required friction is easy; it is more difficult to draw air through a cigarette than through an empty tube because of friction, or rather viscous forces. Air is viscous, though much less than oil, for instance, but oil has difficulty in flowing down a fairly large pipe, so it is easy to see that air will have difficulty in flowing down a very small pipe, which is virtually what it has to do to get between the fibres of the tobacco in the cigarette. If, therefore, a layer or blanket, not of tobacco but of any loose fibrous or cellular material, is placed against the reflecting surface, viscous forces will resist movement of the air particles during compression (and subsequent rarefaction) and thus energy is taken out of the reflected wave. The snag is that if the fibres are too tightly packed, the surface of the blanket gets more solid and we eventually arrive back at square one, with the sound waves being reflected from the surface of the closely packed fibres. It is therefore necessary to compromise, and with most fibrous materials a density of 50 to 200 Kg/m^3 is the most useful range. In the case of cellular materials, it is most important that the cells are interconnecting, for obvious reasons.

Only part of the story of porous absorbers has so far been told because we have yet to consider the all-important factor: the frequency of the sound we are dealing with. Throughout acoustics this is the dominating feature – the effect of frequency. Most sounds, as we have seen, are made up of components of widely differing frequency with wavelengths from twenty or so millimetres to several metres. With low-frequency sounds of several metres wavelength, twenty or thirty millimetres of fibres will not have much effect, but if the thickness is comparable with the wavelength or even greater, the absorber will be most efficient. Thickening the layer can be an expensive matter over a large area, and a most useful improvement can also be obtained by spacing the porous blanket away from the reflecting surface. The result is not that the fibres deal with a longer piece of the wave, as happens when the blanket is made thicker, but that they go to work on a more profitable part of the wave because there is greater movement of air particles with lower frequencies a short distance away from the reflecting surface. The viscous forces in the fibres therefore have more effect.

Odd things start to happen when the wavelength of the sound gets shorter than the thickness of the fibrous layer. The higher the frequency the less distance the air particles travel while being compressed (for a given sound pressure) and are therefore less affected by the viscous forces, but at some frequencies the fibrous layer itself behaves like a spring and the net benefit is reduced. In addition, the effect of the surface of the blanket as a reflector increases with frequency.

When everything is taken into account, obviously frequency is of the greatest importance where absorption efficiency is concerned. Most porous absorbers are poor at low frequency, very good at medium to high frequency and not so good at very high frequency. Anyone who has used an acoustic telephone booth will have noticed that his voice seems to 'boom' more (in the case of a man) because the low-frequency components of his voice are absorbed to a much smaller degree. At best, fibrous blankets absorb anything from 80 per cent to nearly 100 per cent at their most effective frequency, depending on density, porosity and thickness, falling to 60–70 per cent for some acoustic tiles which have to sacrifice efficiency for economy, durability and appearance. The effect of frequency for various types of absorbers can be seen from figure 36. The efficiency of the materials shown is expressed in terms of the absorption coefficient (symbol a). An absorber with 100 per cent efficiency has a coefficient of $1 \cdot 0$, with 50 per cent it has a coefficient of $0 \cdot 5$, and so on.

Compared with solid walls, which reflect about 95 per cent of the sound which strikes them, a wall with a fibrous layer on its surface reflecting only 10 per cent or 20 per cent is very absorbent. Or is it? It pays never to forget the curious way in which loudness and decibels work. Chapter 4 showed that an 80 per cent drop in sound intensity is a change of only 7 dB. Chapter 5 showed that a change of 10 dB is roughly a halving or doubling of loudness. Porous materials are therefore not all that marvellous; if a sound of 80 dB strikes a wall and the reflected wave is 73 dB, it is still pretty loud.

You will no doubt have been thinking that as the viscous forces were working such wonders on the reflected wave they must be doing a grand job on the transmitted wave, the one that goes into or through the wall, and that in other words if you filled up a window with glass wool it would keep out most of the noise. Well, think again! The sound that goes through a layer of fibres has less work

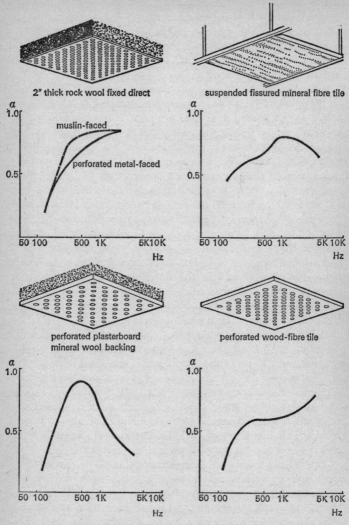

2″ thick rock wool fixed direct

a
1.0

muslin-faced

perforated metal-faced

0.5

50 100   500   1K   5K 10K
Hz

suspended fissured mineral fibre tile

a
1.0

0.5

50 100   500   1K   5K 10K
Hz

perforated plasterboard mineral wool backing

a
1.0

0.5

50 100   500   1K   5K 10K
Hz

perforated wood-fibre tile

a
1.0

0.5

50 100   500   1K   5K 10K
Hz

36. The efficiency of various absorbent materials at different frequencies

done on it than the sound which is reflected. In the latter case the fibres act on the air particles as the compression wave builds up and also as it springs back, but what happens to the sound that springs back has nothing to do with the sound that carries on. The net result is that sound going through a porous blanket only 20 mm or so thick is attenuated by less than 3 decibels, hardly noticeable, and certainly not worth the cost of the glass wool.

An absorption coefficient of 0.8, we have seen, is not startlingly good, but chapter 10 will show that it is quite good enough for many applications. However, there are occasions when something approaching 100 per cent over a wide frequency range is called for – in laboratories and studios, for instance. For every little gain in efficiency over about 80 per cent, the gain in terms of decibel reduction of the reflected wave gets bigger and bigger – another quirk of the decibel scale – and so it is sometimes worth striving for the greatest absorption coefficient possible. How is it done?

The obvious answer is to start by increasing the thickness of the porous blanket. This alone is not enough, and to deal with low-frequency sound would require a ridiculous thickness; the layer of fibres would still present a fairly well-defined surface of their own, to reflect quite a significant amount of high-frequency sound. The solution is to make the fibres into long pointed wedges retained if necessary with open-weave cloth, known as anechoic (an-echo-ic) wedges, whose geometry is such that any sound reflected from one of their surfaces soon hits another and another and so on until it is reduced to an infinitely small intensity. The wedges are often a metre or more deep, with a cavity behind, and have an efficiency as near 100 per cent as makes no difference over a very wide frequency range. Because of their high cost and weird appearance they seldom find their way out of specialist applications.

A cheaper but less efficient type of anechoic treatment consists of the use of layers of fibre or foam of rising density. The outer layer is very light and porous indeed, so that its surface does not do any significant reflecting of its own, but is too loosely packed to provide much of the required viscous forces. Successive layers beneath become denser and denser until they are highly efficient, and the problem of surface reflection that occurs with dense fibres is overcome.

T–E

Still has there been no explanation of the existence of the holes on the ceiling! Neither will curiosity be stemmed by a statement that the holes need not be there at all, and the fibrous material behind would be just as good. Just to be really confusing, let us add that if the fibrous blanket were not there at all, the perforated facing would still absorb sound! If that is not paradoxical enough, how about the statement that neither the holes nor the fibre need be there; just a plain panel would absorb sound? Strangely enough, all these statements are true.

Let us forget about fibres and blankets for the moment and think about holes, or, more precisely, holes with an enclosed body of air behind them. There is of course an everyday object that fits that description: the bottle. Many a schoolboy has played tunes on the cap of his fountain pen by blowing skilfully across the open end and eliciting a whistle. If you extend the experiment to a bottle, the same thing happens, usually with a lower note, and if you take a large bottle with a small neck, like a two-litre wine bottle, and blow carefully enough, a bass tone comes out. No doubt a bottle of sorts was the father of some very early wind instrument; early man may have heard a bottle humming in a breeze thousand of years ago. Man continued to get sounds out of bottles all through history: Vitruvius mentions that sounding vases or echeia were used in Greek open-air theatres, but the idea had to wait until the nineteenth century and the arrival of Hermann V. Helmholtz before anyone looked at it scientifically. Both Helmholtz and his contemporary Lord Rayleigh did valuable work on sound absorption, and their names have been used to describe two types of sound absorber: porous absorbers work on the principle of 'Rayleigh's haystack', and while a bottle is sounding it is behaving as a 'Helmholtz resonator'.

All that a bottle does when you blow across the neck is to resonate, though in a slightly different way from the resonant tube. With this type of resonance it is much more a case of the 'weight-on-a-spring' than of the combination of reflected waves creating an amplified standing wave, which is what happens in a tube. If you block the nozzle of a tyre pump and press on the handle it will behave like a spring. If you fix a weight on top of a spring, push it down and let go, the weight will bounce up and down in a very regular fashion, for the same weight and the same spring, always at the same speed or frequency. For most springs the rate at which they bounce up and down under a given weight or load – the natural frequency – is relatively slow, only a few hundred times a minute. If the load is light and the spring stiff enough, the natural frequency can rise to several hundred times a minute, well within the audible range. Why do springs have a natural frequency? If instead of letting the weight bounce up and down you merely let it settle gently down, the spring will be depressed a finite amount depending not only on the mass of the load, but also on the stiffness of the spring. A stiff spring will not deflect as much as a compliant spring. It also takes a finite time for the spring to settle down under the weight of the load, just as it would take a finite time to spring up again if the load is removed. The frequency at which the spring will oscillate is therefore determined by the distance which the spring deflects and the rate at which it deflects. Exactly the same principles apply to the tyre pump with the nozzle closed as to the ordinary spring.

If the tyre pump is discarded, and bottles are considered again, the same thing happens. If a piston is put in the neck of a bottle, the air in the bottle will be spring-like in just the same way as the tyre pump; for a given bottle the piston will bounce up and down at a rate determined by the piston's weight. Remove the piston and only the weight of the air in the neck of the bottle will be left as a load. The air in the bottle will still behave like a spring, but very much faster because of the very light weight of the plug of air in the neck. It does in fact vibrate fast enough to produce an audible sound, and all that is needed to start it off and keep it going is a disturbance at the neck, which is easily provided by blowing.

The natural frequency of a spring is of course partly governed by stiffness, and also by the mass of the coils of the spring. The same applies to air in a bottle: a small volume of air is stiffer and has less mass than a large one. This is because a given movement of the plug in the neck will cause a proportionately greater compression or rarefaction to a small body of air than to a large one. By filling a bottle gradually with water, the natural frequency can be heard to get higher as the air gets stiffer. The natural frequency of a bottle is also just as much affected by the size of the neck: the larger and shorter the neck, the easier and therefore the faster the air oscillates. Two pint bottles with different-sized necks will produce different frequencies, the one with the smaller neck being lower. The shape of the neck and its length also have an effect, a somewhat complicated one, and the calculating of the exact resonant frequency can be very involved.

What has all this got to do with sound absorption? Bury a bottle in a wall with its neck flush with the wall surface, find out its natural frequency by experiment if necessary and send a sound wave of that frequency towards the wall. If we once more follow the fortunes of the compression wave, it will arrive at the neck of the bottle. Its arrival will be like the push on the weight on a spring: the plug of air in the bottle neck will be pushed inwards. It will soon spring back, overshoot equilibrium position, flow out of the neck. The oscillating plug becomes a sound source itself and sends back or reflects a sound wave which is a replica of the incident wave. Because the bottle is oscillating at the same frequency as the sound wave, the plug will be just starting its return journey at the time when the next compression, one complete cycle behind its predecessor, arrives. The plug, because it is oscillating, has energy which is constantly being converted from potential energy to kinetic energy and back again. This energy, of course, came from the first compression and when the second compression arrives it adds yet more energy, and the plug springs into the bottle with added acceleration.

Now if the sound is suddenly cut off, the plug will stop oscillating after a short time. Why? In theory it should go on for ever, but as with all perpetual-motion machines the trouble is friction, or once again our friend the viscous force. In a pipe, the air next to the walls tends to stick to the walls. Strangely enough, both the pipe wall and the air have surface tension and when the two are in intimate contact all this surface energy is not needed: there is some to spare and this provides the force which keeps the two together. If you blow down the pipe, the air in the middle moves along, but to do so the molecules must slip over their friends who are in such intimate contact with the wall. It is the resistance to this slipping over which causes viscous drag, and absorbs energy (which is again turned into heat). The faster the air travels in the pipe, the greater the viscous drag.

If we return to the bottle, the neck can be considered as just a short length of pipe, with small viscous forces resisting the movement of the plug of air in the neck. It is this effect which stops the plug from oscillating when the sound is cut off. It follows that the viscous forces are in action all the time that the plug is oscillating, so that there is a constant removal of energy which prevents the oscillating building up to an infinite amplitude.

When the sound frequency and the natural frequency of the bottle coincide, at resonance, the air particles of the plug move back and forth very much faster than air particles which are carrying a normal sound wave, and the removal of energy through viscous drag becomes considerable. If you then make the bottle neck much smaller, say by putting a gauze sheet over the neck and thus dividing it into all the little holes in the gauze (the air volume in the bottle will have to be adjusted to maintain the same resonant frequency), clearly the viscous forces become large, so much so that if you cut off the sound the resonance would virtually stop after one oscillation. In other words, by the time the plug springs out of the bottle neck it has lost so much energy that the sound wave it sends back or reflects is very small indeed. We have our sound absorber!

At resonance, the absorption can be almost 100 per cent. The bottle can be forced into oscillating at frequencies near to its natural frequency, but the greater the difference between the two the more reluctant is the bottle to oblige. For this reason a cavity or plain Helmholtz resonator is efficient only at frequencies very close to or at its own natural frequency. Figure 37 illustrates this. The frequency range can be broadened, but the peak efficiency lowered, by filling the bottle with fibres. The effect is to make it easier to resonate at frequencies other than its own, but with much less amplitude: you cannot get a note out of a bottle full of fibres by blowing over the top. Its effect is therefore over a wider range, but its maximum efficiency falls considerably.

At last we come to the holes in the ceiling. They are explained simply by saying that if you put a lot of bottles together you do not

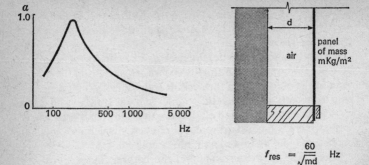

$$f_{res} = \frac{60}{\sqrt{md}} \text{ Hz}$$

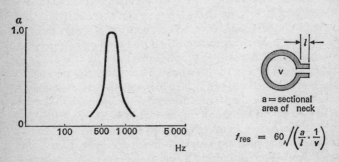

a = sectional
area of neck

$$f_{res} = 60 \Big/ \sqrt{\left(\frac{a}{l} \cdot \frac{1}{v}\right)}$$

37. The absorption of a panel and a Helmholtz resonator

need the walls, merely the necks and the bases, for them still to resonate just as well (figure 38). If the necks are small enough you do not need the gauze to provide the required viscous force. Many acoustic tiles merely amount to a collection of Helmholtz resonators, filled with fibre, without their surplus walls.

This partially explains the paradoxical statements made earlier in the chapter. We have seen how efficient a porous blanket can be at absorbing sound, without any perforated or other facing. In other words, an acoustic ceiling without any holes, just a layer of rock wool, glass wool or suitable foam will do a good job. This is all very well, but how do you keep it up on the ceiling? Fibres have a horrible habit of falling to bits, and glass wool down the back of the neck is not funny. Even in the case of semi-rigid pads and foam

effective shape
of resonator

38. An acoustic tile which acts in the same way as a Helmholtz
resonator

sheets the appearance would be pretty rough, and enough to
frighten an interior designer from ever using sound-absorbing
material.

Fibrous blankets can be held up with mesh to stop them falling
down, but they will not look any better, and if you cover them over
you are liable to lose the much needed porosity. This was a problem
which exercised the minds of the men who first started thinking of
the idea of making commercial acoustic tiles. They would never get
anywhere unless they could produce a durable, pleasing finish.
They set about the problem by starting with the best or only
system available to them: rock wool held in place with chicken
wire. It was the chicken wire they went to work on, trying meshes
with smaller and smaller holes, all the time measuring the sound

absorption, to see how much efficiency they were sacrificing. When they got down to about 30 per cent open area, to their surprise they found that they were sacrificing very little. All they had lost was some high-frequency absorption. They soon found an almost ideal design consisting of a 25mm-thick blanket of rock wool resting in trays of perforated steel sheet which was able to be stove enamelled. The sound-absorbent properties were good and the design was patented.

This was all nearly half a century ago, and now a huge variety of acoustic tiles fills the market. Some are the original design, some use different types of fibre, or a rigid variety with a painted surface subsequently perforated with pin-pricks or larger holes either regular or random. Many use a variety of sizes of holes to obtain a less formal appearance and to stagger the resonant frequencies of the resonators formed by the holes. Some products have slots, because there is no real reason for the holes to be circular.

Their performance varies a little, but all follow the same pattern: poor at low frequency with an improvement when placed over an airspace, good at middle and high frequency, not so good at very high frequency. The holes have been described earlier in the chapter as a collection of resonators, but this effect is only a partial contributor to the efficiency of the tile. At low frequency, the long-wavelength sounds are not so much affected by the holes or the metal between them and the tile is just about as good as it would be without the perforated facing. At middle and high frequencies, less of an area of fibre is presented to the waves, but this loss is made up for by the new effect of the holes as resonators. At very high frequency the efficiency takes a knock, because these frequencies are above the range of the resonator effect and to the short-wavelength sound the area of metal between the holes is significantly reflective.

What about that other statement: that the acoustic ceiling would absorb sound without either fibre or holes? The answer is now fairly simple to understand. A bottle, we know, is a familiar resonator, but there is another equally familiar: the drum. Beat a drum and out comes a musical note, of not very well defined frequency because of multiple resonances, but none the less a note. A drum is really not all that different from a bottle, except that the body of air is enclosed by a flexible skin or membrane, made elastic by being under tension, and all that this does is to introduce more mass and stiffness factors. The mass of the skin is like the load on the spring, and both the tension on the skin and the elasticity of the air in the drum provide the spring-like effect.

Now, if you loosen the drumskin so that it is no longer under tension, the drum will not drum because the limp skin has no wish to be oscillated, its elastic properties no longer being called upon by being under tension. It is now more like the damper discussed earlier in the chapter, except that it is still doing the job of enclosing the body of air. The air is still like a spring and will oscillate, if given enough energy supply to carry its limp passenger the drumskin. Just like the load on the spring, the heavier the skin the lower the natural frequency; and the smaller the air volume the stiffer the air spring and thus the higher the frequency.

If the drum with its limp skin is mounted on a wall, and a sound wave whose frequency coincides with the natural frequency of the drum strikes it, the continuing flow of compression waves provide enough extra energy to make the drum resonate, but in moving and flexing the limp skin a lot of energy is used up, or absorbed. The more limp and plastic the skin, the more it damps the vibration and the more energy it absorbs (yes, it is turned into heat again). Once more, if fibres are put into the drum it will be less dogmatic about the frequency at which it will resonate, and it can be more easily forced into vibration at other near-by frequencies.

A panel or membrane absorber has a performance characteristic very much like that of a cavity or Helmholtz absorber, but at much lower frequency. The natural frequency is a function of the mass of the panel and the depth of the air space. The useful range is from about 40 Hz to 400 Hz, above which it is very difficult to get the skin light enough. A cavity resonator has virtually no upper frequency limit, but space considerations give it a lower limit of around 100 Hz. The panels in a membrane resonator can really be made of any material which fulfils the fundamental requirements: the right mass, enough damping and enough compliance. The first controls the resonant frequency along with the depth of air-space, the second stops the panel from being a sound source itself and does the job of removing the energy, and the third requirement is to ensure that the panel is not too stiff to resonate. Undamped panels with some stiffness can make matters worse and introduce new sounds because of harmonics. This can, however, be turned to advantage if damping is introduced in that they retain these harmonic resonances and give a bit of extra absorption at other, higher frequencies. Of the stiff variety, plywood panels are common, but are rather short of inherent damping. The cutting of grooves in the back will cut down the stiffness and retain most of the mass, and special surface coatings can be used to get added damping. A typical limp membrane can be made with a sheet or two of roofing felt. A fibrous blanket is frequently placed in the airspace behind to broaden the frequency range, and to bring in more damping.

What use are resonant absorbers? They have a limited frequency range, the design factors are critical and because of the amplified oscillation of air, which is the cornerstone of their mechanism, the sound pressure on the back wall of the cavity is increased, with the result that more sound is admitted to the wall. It is the same old story: absorbers absorb sound, and this means less energy in the transmitted wave, but they also help in getting the transmitted wave into the wall, and are thus quite useless at sound insulation.

Resonant absorbers can, however, be most useful; their selectivity can be turned to advantage, for instance to stop a standing wave at a particular frequency in a room or to eliminate an echo in a concert hall. Panel resonators can deal with low-frequency sounds which would otherwise require a great thickness of fibre, and there are a number of tricks which can be used to broaden their range: for example, the back wall of the cavity can be sloped or the hole sizes can be staggered. Cavity resonators can be put to good use in special silencers where much of the noise is at one particular frequency.

9 Sound insulation – the myth exploded

'Soundproofing' is a splendid word. Acoustical engineers are forever being asked to supply 'soundproofing'. After all, waterproof sheeting does not let any water through, so why should there not be special sheeting that will not let any sound through? Once again, the impact of the holes in an acoustic tile is tremendous and confusion between sound absorption and sound insulation is endemic among engineers. It is very understandable because the subject is far from straightforward. The last chapter went to great lengths to show that to erect a thin porous blanket to stop sound going from A to B would be virtually a waste of time. A reduction of 3 dB or so at high frequency would be the most you would get.

Porous materials certainly absorb sound, and because they dissipate some of the sound energy they certainly do reduce the intensity of a sound wave passing through them. However, the thickness of the material must be comparable with the wavelength of the sound for the sound reduction to be worth having, and as we are frequently dealing with sounds having wavelengths measured in metres, porous materials are quite clearly out of the question for use purely as sound insulators. What then can we use? Of what stuff is 'soundproofing' made? We are certainly looking for something fairly remarkable, because a reasonable reduction in sound of say 40 dB means a drop in intensity to one part in ten thousand.

Let us go back and have another look at the various things which occur when a sound wave strikes a surface. In the majority of cases, most of it is reflected, some of it is dissipated and converted into heat, and some of it is admitted by the surface. Because of the law of conservation of energy, in every case the energy in each of these components, the reflected, absorbed and transmitted waves, must add up to an amount equal to that of the wave striking the surface. We are trying to reduce the transmitted wave, and so the logical answer is to increase the reflected or the absorbed wave. We have ruled out the latter course of action on the grounds that it calls for far too bulky and expensive material except for high frequencies, so we are left with only one alternative – increase the intensity of the reflected wave, at the expense of the transmitted wave.

Now the problem becomes much simpler: the more reflective a surface, the less sound is admitted. In the last chapter we discovered what makes a surface reflect sound. With our granite wall, it was so heavy and incompressible that the light air molecules did not have it in them to make much impression on the granite. It would be convenient to be able to have some sort of unit to describe the degree of elasticity and density, and it is fairly easily obtained. You may remember that the speed of sound is a function of the elasticity and density, so this gives us a very convenient method of working out what is called the characteristic impedance of a medium.

The significance of all this is not hard to see. The granite was dense, and because of its relative incompressibility the speed of sound was very high. This makes the characteristic impedance of the granite very high indeed, and the result, we know, was that over 99 per cent of the sound wave was reflected. If, on the other hand, the granite were replaced with simply a 'wall' of air, there would be no transition from low to high impedance, and no reflection. We can therefore say that the greater the mis-match of impedances the greater the reflection of sound and the lower the efficiency of transfer from one medium to the other.

However, 1 per cent or so of sound energy is admitted to the granite, and unless the wall is infinitely thick it will eventually come to the other side and have to start thinking about getting back into air again. Now the granite has a high impedance, which means that compared with air the particles of granite need be barely displaced from their rest positions to transmit high-pressure waves. When the sound wave reaches the far side of the granite, then, the final layer of granite particles will be displaced only a minute amount. Because of the low impedance of air, a given displacement of particles creates much less of a pressure wave in air than in granite. Therefore the arrival of the sound wave at the interface between the two media displaces the air particles by an amount sufficient only to cause a much smaller pressure wave, and so the transfer of energy is again very inefficient.

A thick, dense stone wall is about the best simple sound insulator available. At high frequencies even a 350mm brick wall will give over 60 dB reduction. However, who can afford, spare the space, or allow the weight of several feet of stone for a sound-proof wall? We have really so far been talking about an extreme case. It is far more usual to have a wall which is thin enough for both sides to move almost in phase with one another so that one ceases to be interested in the sound wave travelling in the wall material because the thickness is so short in comparison with the wavelength. In cases like these, the behaviour of the sound wave is different.

Let us imagine a thin panel 3m square, fixed but not clamped at the edges. If a steady pressure is applied to one side of the panel, the amount it will give will be entirely dependent on the stiffness of the panel. If it is a rubber panel it will bulge. If it is a steel panel it will hardly yield. If the pressure on one side starts to alternate slowly from push to pull, the amount that the panel bulges and bows will be governed by its stiffness. However, the panel will have inertia, and as the alternating pressure speeds up there will come a time when the mass of the panel starts to have as much effect as the stiffness. Imagine a door on stiff hinges. If you move it slowly to and fro it will be mostly the stiffness of the hinges which restricts its movement. On the other hand if you start to flap it backwards and forwards fast, it will be the inertia of the door which affects you most and you will be able to flap a plywood door much more easily than an oak one even if its hinges are just as stiff.

At very low frequency, then, we can say that the sound transmission of a panel is stiffness controlled, but as the frequency rises the inertia or mass of the panel starts to become very much more important. Suppose the exciting force is a pure tone. Then the displacement of the panel in the simplest case would be sinusoidal and we are now very familiar with the graph of this function. The rate of change in displacement is another way of describing the velocity, and the rate of change of velocity is of course acceleration.

Figure 39 shows the graphs of displacement, velocity and acceleration for a panel oscillating at two frequencies, one of which is twice the other. For a given sound pressure, particle displacement

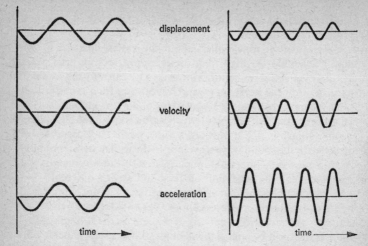

displacement

velocity

acceleration

time ⟶

time ⟶

39. The displacement, velocity and acceleration of a panel oscillating at two frequencies, the second twice the first

decreases with increase in frequency, but particle velocity is constant. The figure shows that for the higher frequency the particle velocity is the same as for the lower, but the rate of change of velocity is twice as high. In other words the peak acceleration is double. According to Newton's Second Law, acceleration is directly proportional to the applied force. In theory, therefore, if the frequency of the sound is doubled, twice as much force would be required to keep the panel oscillating with the same velocity. However, the force applied by the sound wave will be constant, and the result will be that doubling the frequency of the sound will halve the velocity amplitude of the panel.

On the other side of the panel, the effect is exactly the same as if the panel were a sound source. It is being vibrated by the impinging sound behind it, and its forced oscillation emits new sound waves as weaker replicas of the incident waves, on its far side. If the velocity amplitude is halved because the frequency is doubled, you might expect sound pressure to be halved on the far side of the panel. You would be right, in theory. Remembering that a halving of sound pressure is a drop of 6 dB, you can arrive at the statement that a doubling of frequency increases the sound insulation of a panel of given mass by 6 dB.

Newton's Second Law says also that acceleration is inversely proportional to the mass of the body. Therefore, without more ado, let us make another statement, that for a sound of given frequency, the sound insulation of a panel improves by 6 dB for each doubling of mass. Now these are only theories, and we shall soon see that there are many things in practice which prevent a panel conforming to what is known as the 'mass law'. Even at the best of times, the 6 dB improvements turn out to be more like 4 or 5 dB, and so the 'mass law' which one uses in practice is more of an empirical law. It is shown in figure 40.

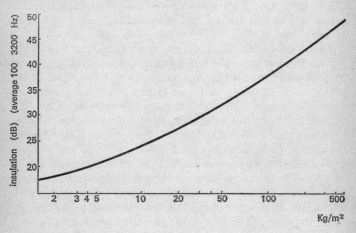

40. The mass law

Now for the pitfalls! You would have some terrible surprises if you designed a sound-insulating panel purely on the basis of the mass law. The panel we were talking about had not only mass, but also stiffness, and although we said that its sound insulation value or transmission loss was only stiffness controlled at very low frequency, this does not mean that stiffness does not come into it at all. Remember the cymbal in chapter 3. It was stiff, and it was because of this that it was able to resonate. A limp cymbal would be useless. If we go back to the panel when it was under steady pressure, it bulged by an amount governed by its stiffness. If, now, the pressure was suddenly cut off, it would spring back. The stiffer the panel, the faster it will spring back, and because it will gain momentum it will overshoot its equilibrium position and in fact oscillate a few times before coming to rest. It will oscillate at a fixed rate, governed by the stiffness and the mass, known as the resonant frequency. Obviously any incident wave which happens to have a frequency which is the same as or close to the resonant frequency of the panel will work constructively with the resonating panel and there will be a very efficient transfer of sound energy from one side of the panel to the other. Add to this the fact that there will be other harmonic frequencies at which the panel will resonate, and you can see that the mass law values would be severely upset.

Just to complicate matters more, there is another type of resonance to deal with. The one we have just described is like the case of the weight on the spring which first appeared in detail in the last chapter. However, when we talked about the cymbal we described its resonance in a different way by talking of bending waves travelling about the surface. Well, a panel is susceptible to bending waves, as shown in figure 41. These waves, which are not the same as sound waves in a solid medium, none the less have the same attributes of frequency and wavelength. If the incident wave strikes the panel obliquely it would be possible for the wavelength of the sound and the wavelength of the bending wave in the panel to coincide. In this case once again the two would work together constructively, and there would be a very efficient transfer of sound energy from one side to the other. You can see from figure 41 that for a given bending wavelength there is a frequency for the incident sound below which the 'coincidence' effect cannot occur because even if the sound wave travels parallel with the panel it will still have a longer wavelength than the bending wave. This is known as the 'critical frequency'.

The result of all this is that one can have a panel which would appear to have adequate mass to be a good sound insulator, but because of resonance and coincidence effects is almost useless at a great many frequencies. What can we do about it? The transmission

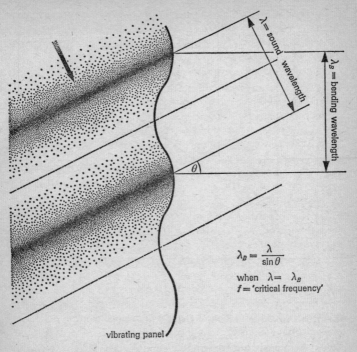

$$\lambda = \text{sound wavelength}$$

$$\lambda_B = \text{bending wavelength}$$

$$\lambda_B = \frac{\lambda}{\sin \theta}$$

when $\lambda = \lambda_B$
$f = \text{'critical frequency'}$

vibrating panel

41. The coincidence effect

loss of this panel would look like figure 42, and it would only be mass-controlled at a small band of frequencies in the middle of the spectrum. One way of tackling the problem is to try to broaden this centre section by pushing the resonances as far down the spectrum and the critical frequency as far up the top end of the spectrum as possible.

The first thing which must go if we pursue this line of attack is stiffness. The softer the spring with a mass fixed to it, the lower its natural frequency. However, not only will this lower the resonant frequencies in the panel, but it will shorten the wavelength of this bending wave which is so troublesome, and therefore raise the critical frequency. Making the panel thinner would also raise the

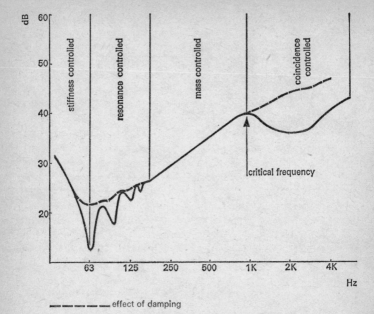

42. Typical sound insulation of a single undamped panel

critical frequency, but we would also lose mass, which we do not want to do. The next most useful thing to do is to increase the mass, not only because of the mass law, but because, provided it is done without any increase in stiffness, it will again lower the resonant frequencies and raise the critical frequency.

There is, however, another line of attack which may be pursued either separately or in conjunction with the one just described. That is to attempt to reduce the amplitude of the resonance and coincidence effects rather than push them to the far ends of the spectrum, where after all we may still need good sound insulation. The reason why resonances make panels useless insulators is, as we have seen, because the panel retains energy from the waves and combines it constructively with subsequent energy arriving with new waves. If we can introduce a means of dissipating energy, the effect of resonance will be greatly reduced.

If you pick up a piece of putty or plastic modelling clay and deform it, you will require a small amount of energy to do it. After you have deformed it and let go, it stays that way. It does not gain any potential energy. What then happened to the energy you put into it by deforming it? The answer is that it was dissipated by internal 'friction' between the molecules of the putty or clay and turned into heat. Now this is just what we want, and if we cover our resonant panel with plastic modelling clay or something similar, each time the panel deforms in the act of resonating it will dissipate energy by deforming the layer of clay as well. What we are doing is introducing damping. Obviously plastic modelling clay is not the thing to use, but there are many substances, from mastic to proprietary compounds, which have the same effect. Going back to figure 42, the dotted lines show how damping can bring our stiff panel back to some sort of usefulness. We can now formulate some general guidelines for the design of panels for sound insulation. For all but very low frequencies, the important facts are:

1. Mass
2. Low stiffness
3. High damping

Now, of course, many substances have a certain degree of internal damping, and the necessity for specially applied means of damping is reduced. Plywood, for instance, has a hundred times more internal damping than aluminium or steel. On the other hand, mass for mass, plywood is much stiffer than steel sheet, and if their damping factors are made up with special treatment such as the application of damping compound, steel sheet is a much better bet because it will have lower resonant frequencies and a higher critical frequency than plywood. Lead sheet is the best of the lot because of its high mass, low stiffness and high internal damping. It is also expensive. Appendix 1 gives some figures for surface density (mass) and critical frequency for various materials. Generally speaking, damped steel sheet is the best compromise between cost, practicability and efficiency as far as single panels go.

However, we have far from reached the end of the road, because there are much more efficient sound insulators than single panels. Suppose we have a damped, low-stiffness panel of, say, $5 \, Kg/m^2$

surface density, and we want to improve on it. What do we do? We probably have already dealt with resonances and coincidence effects to the greatest possible extent, so the only variable left to us is mass. We can increase the mass, but even if we can do this without increasing the stiffness, we cannot hope to get an improvement of more than 6 dB, and in practice only 5 dB or so, because of the mass law. This is hardly just reward for the expense of doubling the materials involved, and, as we might well be looking for as much as a 15 dB improvement, the thought of increasing the surface density to 40 Kg/m^2 is not very encouraging, particularly if we are weight-conscious from the engineering point of view.

Remember the explanation of the characteristic impedance from the early part of the chapter. To make life simple, we discarded it when we started to talk about thin panels, but we should not have done really because whereas we talked about Newton's Second Law and increased acceleration and mass, we were only taking a roundabout way to describe another case of impedance mis-match. What one may call the 'mass reactance' of the panel is a type of impedance, this time called the specific transmission impedance and it is directly related to the mass of the panel and the frequency of the incident sound. Now we can look at the sound insulation mechanism of the panel as being due to a mis-match of impedances between the air and the panel on the incident side and again the panel and the air on the transmission side.

Now if instead of doubling the mass on the panel, we place another identical panel behind the original one with a small airspace between, what happens? We have the equivalent overall mass that we would have had by doubling the thickness of the original panel; but not only this, we have also introduced two more impedance mis-matches because the sound after being transferred into and out of the first panel is now up against the impedance of the second panel, and again there is a further mis-match between the second panel and the air on the far side. One would therefore expect this double panel to be more effective than a single panel of the same total mass. This is certainly true, and in fact if one could separate the two panels completely with a large airspace between, the effect would almost be as good as the arithmetical sum of the decibel transmission loss of two single panels.

Unfortunately, though, in practice the two panels have to be within a short distance of each other, and have some sort of mechanical connexion such as studding. This means that the trapped layer of air combined with the mechanical connexion gives the intervening space a higher impedance than the free air outside, reducing the degree of impedance mis-match within the panel. Nevertheless, double-skin panels are much better than equivalent single-skin ones by as much as 10 dB or more. All the principles of resonance and coincidence still apply, and are in fact much more complicated. A new resonance is introduced, because the trapped layer of air in between the panels can behave like the resonant tube described earlier in the book, and this again goes to decrease the sound insulation of the double panel. Fortunately it is a simple matter to reduce this air resonance by using some of the porous blanket described in the previous chapter. This is a sort of air damper, and has the same effect on resonance in the air cavity as the compound did on the single panel.

One could now go on *ad infinitum* adding more and more separate panels and one would go on getting more and more sound insulation. The all-important factor is the degree of mechanical isolation between the two panels. If they are separated by 'springs' such as a bonded layer of foam rubber, the effect will be very good indeed, except at the resonant frequency of the springs when the amount of damping present will be the governing factor. Unfortunately, high damping in the interspace means high impedance, and as the two panels have high impedances, we are trying to make the intervening impedance as low as possible. Therefore when the two panels are separated by a resilient or spring system, it is better to deal with the spring resonance by getting it as low in frequency as possible, using low stiffness, rather than by using damping.

Sometimes nothing other than a solid mechanical connexion between the two skins is possible. If the airspace between is narrow and the basic wall heavy, the net effect can be little better than for a single panel of the same mass. For instance, a 280mm brick wall with a 50mm cavity and many metal ties bridging the gap is little better than a solid 230mm wall. Removing the ties and any other connexion such as rubble, and filling the cavity with foam, does help a little, but to get any real benefit from double skin construction, an airspace of at least 100mm would be necessary.

The double-skin principle best known to the general public is probably that of the double-glazed window. This also happens to be one of the instances where there is confusion. One thing is certain: two panes of glass both mounted solidly in the frame with only about 20mm separating them are no better at sound insulation than a single pane of the equivalent total mass. An acoustic double-glazed window must have three things:

1. An airspace of 100mm or more.
2. Resilient mounting for at least one pane.
3. Sound-absorbent lining to the reveals.

I am sorry if this hits the sales of conventional thermal double glazing. Happily, the inhabitants of countries like Norway have been accustomed to installing double glazing with a wide airspace for many years and are now delighted to find that the increase in traffic noise is much less of a problem than in other countries.

There are many instances when the multi-skin construction is used for sound insulation. Appendix 2 shows some sound insulation figures for many types of partition or panel. An extreme example is found in studios and laboratories where a concrete room is constructed, and an entirely independent inner concrete room, supported on resilient mountings and with a large airspace all round it, is built inside. The intervening airspace contains sound-absorbent material fixed to the surface, and there are no solid mechanical connexions whatever. The doors have to be specially designed, but we will come to this subject in chapter 13.

Another example is the floating floor. A basic concrete slab has a quilt of rock wool laid upon it, and a 50mm screed, reinforced with wire mesh, is cast on top of the quilt. Providing the quilt is turned up at the edges, and there are no short-circuits by way of skirting boards and the like, this is one of the best ways of obtaining a high sound insulation for a floor. It is also highly effective at insulating the people below from impact noises on the floor above. Floating floors and impact sound insulation will also get special treatment in chapter 13.

Probably most money is wasted in the field of acoustics by house or flat dwellers putting acoustic tiles on their ceiling to try to improve its sound insulation. It does have a small effect, as the next chapter will show, but this is not because it does anything at all to the sound transmission loss through the ceiling. Although in the last chapter we saw that a 50mm-thick porous blanket may reduce the intensity of sound passing through it by 3 dB or so, even this meagre gain is not achieved when the blanket is applied, say, to the ceiling. This is because the porous blanket has an impedance between that of air and the ceiling and it softens the abruptness of the transition from low to high impedance by bridging the two. As sound insulation is dependent in this case upon an abrupt impedance mis-match, the presence of a porous blanket actually admits a higher proportion of sound energy into the ceiling. It does not, however, make matters worse, because it has also dissipated 3 dB or so by means of viscous forces in the interstices between the fibres or the cells. The net result is that the transmission loss of the ceiling is not altered at all.

Let us now make a hole in our sound insulating partition. After all, many of them have to have apertures, even if they are sometimes filled with material of partial efficiency as a sound insulator. Walls have doors, and doors have much less mass than many walls. Worse still, walls sometimes have holes when they should not. What effect do holes or patches of poor insulation have?

We have talked a lot about the transmission loss of a partition in decibels, without really getting down to what this means. There is no difficulty here: if a partition has a transmission loss of 40 dB at a frequency of 1 KHz, obviously the intensity of the transmitted wave will be 40 dB less than that of the incident wave. Now earlier in the chapter we said that the far side of the wall when it is being vibrated by incident sound waves behaves as if it were a sound source. In chapter 4 we took to measuring the total acoustic power output of a sound source, or sound power level, in decibels. There is no reason therefore why the sound which is radiated by the far side of a sound insulating partition cannot be added up to give a sound power level. Sound intensity level, we know, is quoted as a decibel ratio of quantities measured in watts/m^2. Therefore a wall of 4 square metres' surface area will radiate a total power in watts which is four times the intensity in watts/m^2. Of course when adding quantities expressed in decibels one does not add the decibel figures arithmetically, but logarithmically from the table in Appendix 3. We can now say that if our 4 square metre partition lets through a transmitted wave of 50 dB at 100 Hz it will have a

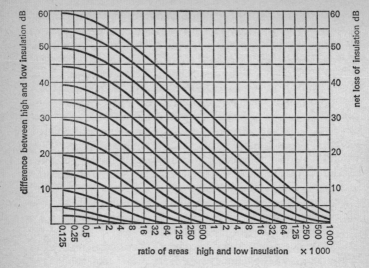

43. Graph showing loss of insulation by composite material with differing areas of high and low insulation

sound power level of 56 dB (doubling a sound intensity is an increase of 3 dB, doubling it again gives another 3 dB).

Now let us make a hole in the partition: what happens? Figure 43 shows a graph from which any permutation of areas of low and high insulation can be entered and the resultant loss of insulation read off.

Let us take another example, a panel with a crack at the bottom. Suppose the crack represents only 1/500th of the area of the partition, which at 4 KHz has a transmission loss of 60 dB. The crack will worsen the efficiency of the panel by 33 dB. It is quite obvious now that it is always essential when good sound insulation is required to make sure that every single crack or gap is sealed up. It is amazing how few people realize this. Later in the book we shall see how important it is to have an airtight seal round a door. Brick walls should always be rendered or plastered to make sure they are not porous.

The detrimental effect of small holes is most noticeable at high frequencies. This is because long wavelength sounds are partially reflected by holes which are small compared with the wavelength. If you are wondering how a hole can reflect sound, remember that the cause of resonance in a tube is the reflection of waves from the open ends. It is a case of impedances again: the impedance of a hole in the wall is higher than that of free air. On top of this, nearly all sound-insulating partitions are much more efficient at high frequency than at low frequency, and therefore a leak of high-frequency sound through a hole is relatively more serious.

Once the sound has got through the hole, there is a difference in the radiation pattern between low and high frequency for reasons of diffraction and interference. High-frequency sound emitting from a hole is very much directional and tends to be beamed to the front. Low-frequency sound is diffracted outwards and has a more or less hemispherical radiation pattern.

Let us summarize what we now know about sound insulation. For a single homogeneous partition:

(1) The surface density should be high, even up to 400 Kg/m^2 for a 230mm brick wall. However, as doubling or halving of mass gives only a 4 or 5 dB change, panels as light as 5 Kg/m^2 have their uses, provided that:

(2) The stiffness is low, unless the partition is required for insulation against sound below 30 Hz or so. Low stiffness pushes resonance and coincidence effects to the extreme ends of the spectrum, where they are often less troublesome. However, resonance and coincidence are markedly reduced if:

(3) The panel has a high damping factor either by virtue of internal damping in the material of which it is made or by having damping material applied to it.

(4) The most efficient partitions for their weight incorporate two or more separate skins separated by as resilient a mechanism as possible. The wider the airspace, the better the transmission loss. Resonance of the intervening airspace reduces the transmission loss at some frequencies (even to the point of making a double-skin partition worse than a single-skin one in extreme cases) so:

(5) The intervening space should contain some form of sound absorption. This may sometimes take on the very efficient double

duty of supporting the second skin as well, as in the case of floating floors and membranes bonded to foam layers.

(6) The whole effect will be ruined by unnecessary cracks, gaps and apertures, but if a local area of lower transmission loss is essential, all is not lost. Figure 43 will show what the net result will be.

(7) Sound absorbent treatment such as a porous blanket is no good for sound insulation alone.

(8) Sound-insulating partitions or panels are nearly always much better at high frequency and sometimes poor at low frequency. This is partly due to the mass law and the 5 dB improvement per doubling of frequency, and partly due to the fact that resonance often occurs at low frequency. Single-figure averages in decibels for the sound insulation of partitions can therefore be very misleading, and, as always, it is most important to make the calculations for the complete range of frequencies of interest. Octave bands (chapter 4) are most convenient for this. Averages in decibels over the range 100–3,200 Hz are, however, sometimes used.

10 Rooms, buildings and concert halls – flying saucers in the roof

If you are one of those joyful creatures who makes a habit of singing in the bath, you will be well aware that sound in a room sometimes behaves in a very different way from sound out of doors. Motorists know that as they drive their cars out of the garage there is a sudden drop in noise level as it reaches the open air, and anyone who has moved house will know how strangely sound behaves after the furniture has gone.

The reason for these phenomena is no great mystery, but it is in fact rather more interesting than at first appears. The study of sound in rooms is also of as much importance as any other aspect of acoustics, not only because the enjoyment of thousands of music lovers depends on architects' knowledge of the subject, but perhaps even more because many noises are made and heard indoors and are injuring workpeople's hearing, often because of the effect of the surroundings. This chapter brings together the subjects of the previous chapters, and shows how all the phenomena of radiation, diffraction, absorption and reflection combine to make life pleasant or unpleasant for us room dwellers.

Let us start with a small empty room, one that has dimensions only a few times the wavelength of the sound we are interested in. If we place a small non-directional pure tone source in the centre, sound waves will radiate in all directions and very soon strike the walls of the room. In a room like a bathroom, the hard solid walls will reflect over 95 per cent of the sound energy, and so the waves will be almost completely reflected, as light is from a mirror. Now we know that the combination of an incident wave and its reflection produces a standing wave pattern; therefore there will be a pattern of standing waves extending away from the four walls. The length of each wave will be equal to the wavelength of the incident sound. If, however, the distance between any of the three pairs of parallel surfaces of the room is equal to or an integral multiple of the wave-length of the sound, the standing wave patterns will coincide and combine constructively to form yet another type of resonance. Really all that we have is a three-dimensional resonant tube.

Unfortunately the pattern of standing waves in a small room is very complex because of all the permutations of length, breadth and height which determine at what frequencies resonance occurs. To work out all the frequencies at which these resonances, which are properly called 'normal modes' or 'eigentones', occur takes a very long time without the aid of a computer. The subjective effect of these eigentones is twofold: firstly, in the presence of a constant sound one can move about the room and enter pockets of high and low intensity, but secondly, and bath-time baritones will be familiar with this, certain frequencies get preferential treatment and are singled out and amplified. The most lovely rich note can be sung in a bathroom which would sound quite watery in the bedroom.

In a larger room, except at low frequency, the number of eigentones increases so much and they are distributed over so many different frequencies that they can be ignored, and we have what we call a diffuse sound field. The sound from the source is reflected over and over again from the walls of the room, but at all except low frequencies there is no important standing wave pattern. It is now much easier to work out what effect the presence of the room will have on the sound level of the source.

Let us consider a spherical source of 100 dB sound power level. In the open air, the inverse-square law would tell us that the sound intensity at 3m distance would be 80 dB. Now let us take this source into a large room, say 10m × 10m × 3m high. Let us assume that the absorption coefficient of the walls, ceiling and floor of this room was 0·05, as it would be if they were plastered brick or concrete. What would we hear? First of all there would still be the direct sound coming straight from the source to our ear, and as nothing has been done to the source itself or been placed between it and our ears, it will still be 80 dB. However after we have heard the direct sound, it will go speeding past our ears and strike the walls, ceiling and floor. These will absorb 5 per cent, and reflect 95 per cent of the sound energy back to us. Once again the reflected waves will go speeding past us and the same thing will happen. In fact, the sound will have to be reflected over four times before even 20 per cent or 1 dB has been removed from it. The intensity of the first reflected wave will be increased 18 times by the adding in of all the subsequent reflections as they slowly die away to an insignificant level. If you add up the intensities of all the reflections the answer will be an increase by a factor of

$$\frac{1 - \bar{\alpha}}{\bar{\alpha}}$$

where $\bar{\alpha}$ is the average absorption coefficient of the wall. This quantity is not difficult to calculate. Different materials absorb varying amounts of sound, and their absorption coefficients at different frequencies can be taken from the results of measurements. Appendix 4 gives absorption coefficients for many materials. The total absorption of a particular patch of material is the product of its absorption coefficient and area, and the total absorption in a room is the sum total of absorption of all the different patches of material. The average absorption coefficient is obtained merely by dividing by the total surface area. In very large rooms, such as large concert halls, the viscosity of air itself causes energy dissipation at high frequency of a sufficient degree to require it to be taken into account. This is known as air absorption, and when it is included in the average absorption coefficient the symbol $\bar{\alpha}_T$ is used.

If then we calculate the intensity of the first reflected wave and multiply it 18 times, we will know the overall sound intensity which is due to reflection. As this must hold good for any position in the room, we have to work out an average value. Suppose the room were a sphere; the intensity of a sound from a source in the centre would at the surface of the room be $\dfrac{1}{4\pi r^2}$ of the total sound energy of the source. If the walls were totally reflective, the reflected wave would converge back on the centre and, as 'r' would there be 0, the intensity of the reflected wave would be equal to the sound power of the source. However, we are interested in the average level of the reflected wave, not the level at the extremity or the centre of this room. We will get the right answer if we work out the intensity at a point halfway from the centre to the surface, that is

$$\frac{1}{4\pi(\frac{r}{2})^2} \quad = \quad 4\left(\frac{1}{4\pi r^2}\right)$$

This would also be the case if the room was rectangular; the average level of the first reflected wave for a totally reflective room would be $\frac{4}{S}$ of the sound power, where 'S' is the surface area of the room. We can now say that the total level of the reflected sound will be

$$\text{SWL} + 10\log_{10}\frac{4(1-\bar{\alpha})}{S\bar{\alpha}}$$

where SWL is the sound power level.

For convenience, $\frac{S}{1-\alpha}$ is called the 'room constant' or 'R', so the formula becomes

$$\text{SPL} = \text{SWL} + 10\log_{10}\frac{4}{R}$$

Figure 44 shows a graph of this formula, so that the reverberant sound pressure level can be read off if one knows the total surface area and total absorption in the room. Some important factors emerge:

(1) Throughout most of the range a doubling of $\bar{\alpha}$ gives a 3 dB drop in reverberant sound level.
(2) For the same value of $\bar{\alpha}$ the larger the surface area, which usually means the larger the room, the lower will be the reverberant sound level.

In our example, the sound power level was 100 dB, the surface area of the room was 320 square metres, and the average absorption

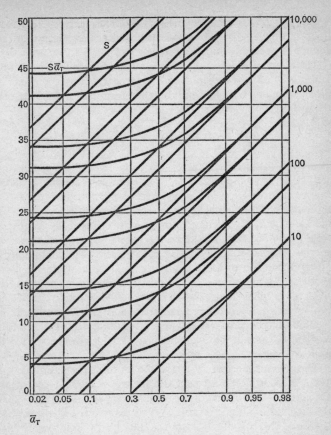

44. Graph for obtaining reverberant sound pressure level from sound power level of source

coefficient was 0·05. Insert these values into the formula (or figure 44), and we have

$$SPL = SWL + 10 \log_{10} \frac{4 (1 - 0·05)}{320 \times 0·05}$$

$$= 100 + 10 \log 0·25 \text{ dB}$$

$$= 100 - 6 = 94 \text{ dB}$$

When you remember that this same source gives rise to direct noise of only 80 dB at three metres distance, the presence of the room has pushed the noise up somewhat!

The effect of reflections in a room is known as reverberation, and our 94 dB is reverberant sound. In nearly every case there is direct sound as well, and the complete formula for working out the overall sound level is

$$\text{SPL} = \text{SWL} + 10 \log_{10} \frac{Q}{4\pi r^2} + \frac{4}{R}$$

where Q is the directivity factor of the source, 'r' is the distance of the listener from the source, and R is the room constant.

From figure 44 you can see an interesting point: if $\bar{\alpha}$ is 0, that is if the room is totally reflective, the sound intensity in the room is infinitely high. If you come to think of it, this is really what you would expect; there would be the sound source pumping out energy which bounced around forever and was continuously piling up on top of more and more sound energy. In practice, though, no surface is totally reflective. Another thing you can see from the graph is that as the average sound absorption coefficient $\bar{\alpha}$ increases, the level of reverberant sound starts to fall considerably. If in our 320m² room $\bar{\alpha}$ is increased from 0·05 to 0·1, there is a 3 dB improvement. If we can get it up to 0·5 there would be a 13 dB improvement.

Throughout most of the range the moral is clear: double the average absorption coefficient and there is a 3 dB improvement. However, at the top end of the range, where you might think that the law of diminishing returns would start to operate, the reverse happens. This sort of absorption would only occur in very special circumstances indeed, such as an anechoic chamber, but an improvement from 0·95 to 0·98, only another 3 per cent, gives a 4 dB improvement.

Going back to our reverberant room with 94 dB, if we were to stand inside it we would have to get within 300mm of the source before the direct sound would rise even to the level of the reverberant sound. On the other hand, if the sound absorption of the room surfaces is increased, the level of reverberant sound decreases, and direct sound predominates for a greater distance from the source. Reverberant sound is more or less uniform in level throughout the room, whereas direct sound falls off with increasing distance.

If the sound source is now suddenly cut off, what happens? Instant silence? No, because it takes a certain amount of time for the last few sound waves which the source emitted to hit enough surfaces and be completely absorbed. If you clap your hands in a reverberant room, you can actually hear the clap die away as its sound waves shuttle backwards and forwards from surface to surface. The time taken for a sound to decay by 60 dB is called the 'reverberation time' of the room. In laboratory conditions it is possible to obtain a reverberation time of as much as 15 seconds. Anyone who has visited the Baptistery at Pisa, next door to the famous leaning tower, may have been lucky enough to get the attendant to sing an arpeggio. The reverberation time in the Baptistery is about 12 seconds, and consequently the chord takes nearly a quarter of a minute to die away, during which time the 'timbre' or quality of the sound alters in a remarkable way as the low-frequency components and fundamentals die away first leaving the colourful harmonics to linger on. Even for the non-musical, the sheer volume of sound is impressive.

It is not difficult to see that the reverberation time of a room which produces a diffuse sound field will be dependent on the absorption coefficient of the surfaces. The marble dome and walls of the Baptistery give it an average absorption coefficient of 0·03. However, we are primarily interested in the time the sound takes to decay and this is also dependent on the room volume.

The scientist Sabine assumed that the decay of sound was continuous, and produced the formula:

$$T = \frac{0·16V}{S\bar{\alpha}}$$

where V = room volume. However, if the average absorption coefficient is high, the decay will not be continuous, but will be 'stepped' as each reflection occurs and a large 'chunk' of energy is absorbed. Eyring modified the formula as follows:

$$T = \frac{0·16V}{-S\log_e(1-\bar{\alpha})}$$

to take account of this, and Knudsen modified it yet again to include air absorption by adding the term '$4mV$' to the denominator to represent air absorption.

The interesting thing now is that whereas the steady-state reverberant sound level is lower in a large room than a small room with the same average absorption coefficient, the complete opposite occurs with reverberation time. Because in a small room the sound waves have a much shorter distance to travel between reflections, they are absorbed much faster than in a large room, where the same number of reflections take much longer to occur.

Reverberation time is seldom of direct interest in noise problems, but its measurement provides a valuable means of determining the total absorption in a room. Total absorption and total surface area are the only quantities which affect the steady-state reverberant sound level, which is the factor that is often of the greatest importance. Reverberation time is normally measured using a microphone, amplifier and graphic level recorder in which a high-speed stylus describes the decay pattern on a moving roll of graph paper.

It is when one comes to design auditoria that reverberation time is of paramount importance, and, although this book is not really concerned with this aspect of acoustics, it is interesting to take a brief look at the subject. What are 'good acoustics' anyway? Concert-goers are very familiar with the fact that some halls are better than others, but few know why. There are, in fact, three separate conditions which all have to be fulfilled for a concert hall to have good acoustics. First of all, the reverberation time must be right. If it is too long, the room will be too reverberant; successive musical notes will run into one another and the effect will have something in common with the playing of a piano with the sustaining pedal forever in use. In fact, church music has evolved as it has in order that it can be played or sung in places with long reverberation times. Plainsong is ideal because there are few changes of pitch, but when they do occur the lingering on of the previous notes enhances the effect. However, an orchestral concert in a church would be an awful jumble of discordant sounds. Equally so, a band recital in Hyde Park tends to sound rather 'thin' and unbalanced because of the lack of any reinforcement by reverberation. Figure 45 shows some optimum values of reverberation

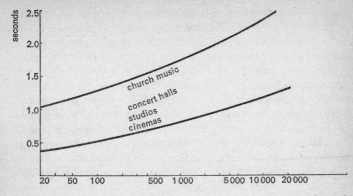

45. Typical reverberation times

time at 500 Hz for various types of auditoria or buildings. For speech only, a short reverberation time is required to ensure maximum intelligibility.

However, everything can be ruined by the presence of echoes. What is an echo if it is not reverberation? Once again it is a question of time. When a sound is made in an auditorium, the first thing a member of the audience hears is the direct wave. This will be followed immediately by the first reflected waves. Haas and Meyer in Germany conducted experiments which showed that providing the reflected waves arrived at the listener within 35milliseconds of the direct sound, he would 'hear' only one single sound as if it were coming straight from the source. On the other hand, if the reflected sound took longer than 50milliseconds, it would be heard as a distinct echo in its own right. Bearing in mind the velocity of sound, 334m/sec, this means that any reflected path over about 12m long is liable to create an echo. The remedy is merely a matter of the application of sound-absorbent treatment to such surfaces that could cause this. Another type of 'echo' can be caused by resonances, such as the eigentones described earlier, which result from the interaction of standing waves between parallel surfaces. This causes two things: firstly, certain musical notes are excessively amplified, and, secondly, the reverberation time at these frequencies is lengthened to result in particular notes lingering on after the rest of the sound has died down.

The third aspect of a good auditorium is its geometry. It must be designed not only to be free of any configurations which may cause echoes or resonances, but also so that every member of the audience can hear well. This can be difficult, because the means of achieving it involve utilizing wall and ceiling surfaces to reflect sound to all of the seats, and making sure that these reflections arrive within 35 milliseconds. A major problem can arise, in that it is often necessary to have a large volume, and thus a high ceiling, to achieve the required reverberation time, and yet reflections from a high ceiling would arrive too late. The answer is to use suspended reflectors at the ideal height but with plenty of spaces between them to admit sound to the space above. Many modern halls incorporate this system.

The Royal Albert Hall used to be infamous for the echo which resulted from the high domed roof. Not only was there a very long path for the reflected wave, but the dome also had a focusing effect, causing amplification of the echo. Attempts to ensure that first reflections occurred earlier resulted in the construction of a canopy over the orchestra, but the echo could be eliminated only by installing porous sound-absorbent 'flying saucers' in the dome.

Sometimes it can be quite a problem to achieve an adequately long reverberation time in a hall, particularly one of small volume per seat. This is because people and seats absorb sound; the total absorption of a seated member of the audience is 5·0 at 500 Hz, and consequently the audience can provide 50 per cent of the total absorption present. No modern opera-goer is able to enjoy the acoustics of famous old halls quite as their grandparents did. Styles of dress have changed, so much that the absorption of the audience has gone down quite considerably. The sound absorption of bare thighs and a mini-skirt cannot compete with layers of voluminous petticoats, with the result that the reverberation times are undoubtedly longer now than they were a few generations ago.

Often it is necessary either to test halls or to rehearse orchestras when there is no audience present. This will of course lengthen the reverberation time and alter the acoustics of the hall. To overcome this, seats often have perforated undersides to form Helmholtz resonators, so that when not in use and folded up the resonators come into play and compensate somewhat for the absence of the occupant. A fully upholstered, unoccupied seat has a total absorption of 3·0 at 500 Hz. An experiment was conducted in America where expanded polyurethane blocks were seated in place of the audience at a test recital in a new hall. This also ensured that the orchestra was safe from a hostile reception!

Normally, though, when sound absorption is called for, middle- and high-frequency treatment is achieved with combinations of porous blankets and Helmholtz resonators, and low-frequency absorption is obtained from membrane or panel resonators. Large Helmholtz resonators are occasionally used at low frequency also.

In all our calculations of reverberation so far, we have been dealing only with rooms having a diffuse sound field, and this applies by no means to every room. To begin with, in a rectangular room the reverberation time can be complicated by the presence of an acoustic ceiling. Sound waves reflected to and from the ceiling and the floor will be attenuated much more swiftly than those which shuttle back and forth between reflective walls. The result is an initial period when the sound decays quickly, followed by a period of slower decay. In these cases the ordinary formula can be misleading for calculation of an absolute level of reverberant sound resulting from a source of known sound power level. It is still useful for calculating the relative effect of increasing the sound absorption in the room.

All formulae go haywire when one comes to consider the latest development in acoustics, the open-plan or landscaped office. Acousticians are only just beginning to appreciate what happens to sound in an environment when the lateral dimensions of the room are often over ten times the height, and when both the ceiling and floor are sound-absorbent. In a conventional diffuse field, calculations rely on sound striking all parts of the surface very regularly, with the result that the level of reverberant sound is almost constant throughout the room. In a landscaped office conditions are very different, and no reverberant field in the conventional sense of the word exists.

Direct sound is still inversely proportional to the square of the distance. As far as reverberation is concerned, one can ignore the effect of the walls and treat the room as infinitely long and wide, in which case the only dimension of interest will be the height. Figure 46 shows a source and a receiving point in a typical section of a

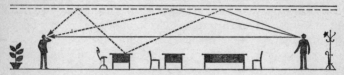

46. Sound paths in a landscaped office

landscaped office. From this you can see that the strongest reflected waves will be the ones which have been reflected once only from the floor or the ceiling. The floor will nearly always be more reflective than the ceiling. The strength of these waves will be governed by two things: the length of the path, according to the inverse-square law, and the absorption of the ceiling or floor. If we know both these factors we can work out the intensity of the first reflected wave.

However, we have got to add to this the next strongest waves, the ones which have been reflected once from the floor *and* once from the ceiling, and after this come the waves which have twice hit the ceiling and once the floor, or vice-versa. In fact, there is an infinite number of paths, but soon we reach the point where the sound has been reflected so many times that it is insignificant.

Up to 10m or so from the source, this distance 'as the crow flies' is much shorter than the distance the reflected waves travel, as figure 46 shows. This means that if you double the direct distance, you do not double the reflected path. Until you get 30m or so from the source, the effect of absorption by the ceiling and floor is more or less constant. What happens after that we shall soon see.

What it boils down to is this: close to the source, direct sound predominates over reflected sound and the latter can be ignored. If the sound-power level of the source is known, the sound-pressure level close to the source can be worked out using the inverse-square law, and it will fall by 6 dB every time the distance is doubled. If the average absorption coefficient is greater than about 0·5, direct sound will predominate everywhere. However, the poor absorption of the floor, which may be covered with desks and furniture, will give a value in most cases of 0·4 or thereabouts at 15 Hz and above, and as little as 0·2 below. In these cases, there comes a point when the reflected waves, which are falling off at a rate of only about 3 dB per doubling of distance, exceed the direct waves in intensity. This point usually occurs at about 3m from the source, and extends up to 10m or so, when the reflected path becomes relatively close in length to the direct path, and the 6 dB drop per doubling of distance starts to apply again.

One important consequence is that an occupant in a landscaped office does not get any subjective effect of reverberation, and all sound appears to be direct sound. This is because the reflected waves always arrive within the critical 35milliseconds period referred to earlier, except in the far field. In practice, landscaped offices of course contain not one but a multitude of sources. The calculation of the ambient sound pressure level is therefore not very straightforward. It involves knowledge of the location of the sources, and a statistical estimate of the distribution and number of sources actually emitting sound at any one moment. In most cases one can ignore the far-field type of reverberation because unless a part of the office is used for much noisier activities than another, the first type of reverberation due to local sources will predominate.

Here we must leave landscaped offices, because for the moment we are studying only the behaviour of sound, and not criteria and means of bringing it about. We will be coming back to the subject later.

Let us now look at a few of the implications of what we have discovered about sound in rooms. The first is that sound behaves in the same way as it does out of doors only in rooms with walls which are virtually totally absorbent to sound. In all other cases there will be two, if not three, fields of noise. In conventional rooms there will be the direct field where direct noise is sufficiently greater in intensity to be the dominating factor. Within this field, the sound is directional from the listener's point of view, and subject to the inverse-square law. However, when there are a considerable number of sources in a reflective room, the combined reverberant field can exceed the individual direct fields in intensity up to within very short distances from a source, or even totally.

Then there will be the reverberant field, which becomes the major factor a certain distance from every source. In this area, the sound engulfs the listener from all directions and the subjective effect is totally different. Because of the lack of directivity in the reverberant field, noise control by screening is obviously useless. In open-plan offices, with very large lateral dimensions and an acoustic ceiling, the most important sound field occurs between the direct field and the true reverberant field, which may be called the indirect field.

We have so far tacitly assumed that the sound source will always be inside the room in question. Of course this is nonsense, and anyone who has winced as a jet airliner passed low over his roof will hasten to agree. What sort of sound field have we in this case? The answer is much simpler than it first seems. In many cases the sound from the external source will enter the room by penetrating a wall. We have already seen that a wall that is being excited by an incident sound wave or waves is virtually a sound source itself on the transmission side. Therefore if sound penetrates a wall, the wall becomes a sound source as far as the room the other side is concerned, and its sound power is a function of the intensity of the transmitted wave and the area of the wall. Bringing in the decibels again, one discovers that a 10m² wall will emit 3 dB more sound energy than a 5m² wall. At very low frequency, the wall vibrates in such a way as to emit plane waves, like the ones caused by the piston in the tube, which are not affected by distance. However, at most frequencies of interest walls do not vibrate in 'unison' and their mode of vibration is very complicated. In all but the most highly absorbent rooms, the sound field in the room into which the sound is transmitted is predominantly reverberant.

This brings to light a very important point: the effectiveness of the walls of a room at sound insulation depends not only on the transmission loss of the walls, but also on the absorption of the room. When the sound is entering principally via a relatively small area like a window, however, the direct sound field can become important for some distance from the window, and the absorption in the room will of course affect only the reverberant field.

Now let us take the reverse situation: suppose we were interested in the sound pressure level at the point outside. What difference does the presence of the enclosure make? The reverberant sound at the boundary of the room will effectively increase the sound level of the source, but this will be offset by the transmission loss of the walls. In fact the direct sound will be reinforced by reflected waves inside the enclosure by a factor of $\frac{1}{\bar{\alpha}}$. The walls of the enclosure can be regarded as a new sound source, with a sound power level equal to

$$SWL_1 - TL + 10 \log \bar{\alpha}$$

The sound reduction of the enclosure will therefore be equal to

$$TL + 10 \log \bar{\alpha} \text{ dB}$$

This means that if $\bar{\alpha}$ is only small, the transmission loss of the enclosure will be severely impaired. If $\bar{\alpha}$ is 0·05, for instance, the value of the enclosure at sound insulation will be cut down by as much as 13 dB.

What then do all these figures amount to? The biggest lesson to learn from them is that it is no good simply subtracting the transmission loss value of a wall from the original sound pressure level and hoping that the answer will give the sound pressure level after building a room, whether to enclose, exclude or partition off a source. One must take into account the area of the wall and the absorption in the enclosure or room. Many people have discovered to their cost that it is possible to box in a noise source with precious little improvement because of internal reflectiveness.

There is one more large snag. Everything that has been said so far in this chapter has been about airborne sound. What has not been said is that whenever sound enters something like a brick wall it immediately becomes solid-borne sound, even if it transmits the other side as airborne sound again. Sound in solids travels great distances with only small losses. It is not subject to the inverse-

47. Transmission paths between rooms

square law when it is being channelled along a wall because it then consists of either bending or plane waves, not spherical waves. This can have very unfortunate results. Sound in one room impinges on the walls and becomes solid-borne sound; even if the wall admits only 10 per cent of the airborne sound because of the impedance mis-match, there is only a 10 dB drop in intensity. Once in the solid wall, the solid-borne sound waves will not only create airborne sound waves on the other side, but will travel along the wall and into the rest of the building structure. The result is that if you have two or three adjacent rooms in a building, the path for sound from one to another will not only be by direct penetration of the walls, but also by travelling along the walls and the floor and then re-radiating as airborne noise. Figure 47 shows all the paths that sound can take between rooms. Unfortunately concrete is one of the most efficient conductors of solid-borne sound, and because it has very poor internal damping, sound will travel from the top of a block of flats to the bottom by way of the concrete.

All our formulae can go by the board if solid-borne sound is a major factor, and this means that we have got to be very careful indeed what figures we use for transmission loss. The worst cases where 'flanking' transmission, as it is called, occurs are in sound transmission between adjacent rooms. For this reason the theoretical transmission loss of the party wall would give us very over-optimistic results, and it is necessary to rely on values for different types of wall which have been measured in field tests rather than laboratory tests, and certainly not to use the basic mass law. Solid-borne sound of course can sometimes start off life in the solid and not originate as airborne sound. This occurs when machinery is fixed direct to the structure, or when feet fall on floors, as they are wont to do. Chapter 13 will look at methods of dealing with these cases.

11 How quiet should it be? – comfort, duty, and the law

'Silencing' is just about as misleading a word as 'soundproofing'. Both words suggest total exclusion or eradication of noise, and this is almost never possible or necessary. For one thing, regardless of the initial sound level, to reduce it to zero sound pressure would call for a reduction of an infinite number of decibels.* Some specifications are stiff, but not even the most impractical designer or engineer would call for this!

What, then, should he call for? How much noise annoys? How much is safe? What does the law require? If we can answer these questions, we can answer the all-important question: 'How much will noise control cost?' All noise control costs money, and to make matters worse it often does not warrant being described as an 'investment' because a return on the capital involved is hard to find. All too often it is a palliative forced upon management by protesting neighbours or suffering workpeople. 'Over-silencing' is money down the drain. One of the most valuable contributions the acoustician can make is an exact assessment of the right compromise between effectiveness and economy. 'Under-silencing' is just as wasteful, because noise control as an afterthought, or measures which are not good enough and have to be improved on, are always vastly more costly than the correct treatment would have been in the first place.

On the brighter side, there are certainly situations where noise control is an investment as a means of increasing productivity. Few will dispute, for instance, that office staff will work better in an acoustically well-designed office than in a very noisy one. In landscaped offices, for instance, the combined effort which goes

*$20 \log_{10} \dfrac{p}{0} = \infty$

into the whole question of environmental design produces much greater efficiency than in the conventional open-plan office. In factories there is evidence that accident rates and absenteeism are reduced, staff turnover is lower, and improved efficiency results. There was a case a few years ago of a photographic factory in which was conducted a carefully controlled experiment to determine whether the rate of film breakages as a result of machine mismanagement on the part of the operators would be reduced if the noise level was cut down. It was found that men working in rooms which had been acoustically treated worked no faster than those in untreated rooms, but the rate of film breakage was reduced to one fifth of its previous rate. In another well-known case, weavers given earplugs were found to become about 12 per cent more efficient than previously.

However, efficiency is by no means the prime concern in noise problems. In the great majority of cases it is a question of avoiding annoying someone which is behind the need for noise control, and as far as annoyance goes it very often takes place in people's homes. What then constitutes an annoying noise to a non-neurotic person with unimpaired hearing?

In nearly all cases of annoyance, it is not the absolute level of the noise which is the important factor. The most important thing to ascertain is the 'noise climate' of the environment. It was once said in court that what would be a nuisance in Belgrave Square would not necessarily be so in Bermondsey! Although this might seem a rather 'right-wing' remark, it is true for acoustical, not purely social, reasons. It is not that those who have to foot astronomical rates and rents are jolly well entitled to a bit of peace and quiet, but that the 'noise-climate' in Belgrave Square is totally different from that in Bermondsey.

The noise from a few fans on a chicken farm can constitute a major problem in deep country, but would be totally ignored in a suburban housing estate. Many an iron-foundry pollutes the surrounding area with noises which would be intolerable in most other areas. As it happens it may well be quite tolerable in its own environment because it will be completely in character with the surroundings, have other noisy works near it, and to such residents as there may be it probably represents their living. Coalminers may be indifferent to having to look at the slurry of a slag heap every day because they have helped put it there, and doing so has been their source of income. Put the same slag heap in, say, Welwyn Garden City and see what happens.

It is therefore not possible to say something like '35 dBA is an acceptable noise level in bedrooms'. There are plenty of cases where it would be unnecessarily low. On top of this, one noise of 35 dBA may be very much more annoying than another of the same level. There has been enough about noise in this book to make it plain that two totally different sounds can both turn out to be the same in dBA.

The simplest case of this state of affairs occurs when a noise which consists of one or a few pure tones is compared with one which has no character whatever, such as wind noise. A typical noise polluter is a cooling tower. These structures often take the form of a timber tower with a large axial fan on the top; air is cooled by being drawn up through a cascade of water. Often the fan runs so slowly that no blade-pass note is audible – the noise is mainly aerodynamic and due to the falling water. Even if it has considerable intensity at low frequency, it is a generally tone-free noise. Now take another fan, an exhaust fan on a building, which may run at several times the speed of the cooling-tower fan and have a large concentration of energy in one or two discrete tones. Although the sound level at a near-by house from each of them may be the same, say 40 dBA, the exhaust fan will probably constitute a noise problem, whereas the cooling tower will not, and may even be beneficial.

Beneficial? This raises another point: sometimes the answer to a noise problem is not the reduction of noise, but the introduction of noise. The cooling tower might exist on a factory site when there were continuous comings and goings, clatters and bangs, and even if the noise of these is reasonably quiet across the road, it can be very irritating. The presence of the cooling tower provides what is known as a mask. In itself it has no seriously annoying attributes, and because of the ear's inability to hear some noises in the presence of others, as we saw in chapter 5, it removes the nuisance value of the clatters and bangs, and possibly a few screaming fans, if they are not too loud.

How then do you go about assessing a noise problem? How do you decide how much of a nuisance it is? For the man who has only a simple sound-level meter, perhaps the best thing is to turn to an existing British Standard, No. 4142: 1967, entitled 'Method of Rating Industrial Noise Affecting Mixed Residential and Industrial Areas'. Although the International Standards Organisation may be going to recommend lower figures than the British Standard, in many ways it is most useful.

First of all it allows for two types of situation: the case when the offending noise can be temporarily turned off so that a measurement of the background noise level can be obtained, and the case when it cannot. The offending noise level is measured in dBA at a point outside the building where complaints do or may originate – at a height of 1·2m above ground and at least 3·6m away from any walls. It is important to make sure that noise due to wind, electrical interference or other alien sources is eliminated, and even with a wind shield on the microphone it is not advisable to take measurements when there is more than a light breeze. In the case of very low readings remember that every amplifier circuit has a certain amount of internal noise which will provide the lower limit on the range of the meter.

In many cases the noise is by no means steady – the level may fluctuate considerably. One then tries to obtain two figures, the first being a visual average of the needle fluctuations, if these are within about 10 dBA, the second being the level which the needle may from time to time shoot up to. The measurements should continue over a long enough period of time for a representative estimate to be made of the occurrence and duration of the two levels, which will be known as L_1 and L_2.

Then one must take into account the character of the noise. If it is irregular and impulsive, such as the crashes and bangs referred to earlier, then 5 dBA must be added to the measured level to compensate for the extra nuisance value. The same must be done if the noise has a definite distinguishable continuous note, even if this is not a pure tone but has a definite pitch. However, these are not the only attributes a noise may have which will affect its annoyance value. An intermittent noise is considered to be less annoying than one which continues all the time, although if the noise regularly cuts in and out it will become more troublesome because each time it either cuts in or out it draws attention to itself. You have probably had the experience of noticing that a clock has stopped or the heating has cut out, although while they were running you did not hear them because they were so familiar to you. It was the sudden absence of sound that caught your attention.

However, if the total off-time is more than a few per cent of the total time, the overall annoyance value is reduced. A typical case, although not one covered by the British Standard, is that of aircraft noise. Anyone living in the vicinity of Heathrow will agree that life would be much more pleasant the same distance from Gatwick Airport, where the off-time percentage is much higher. A correction is therefore applied to the noise level to take the intermittency of the noise into account. Graphs are published in the Standard to correlate the typical on-time duration, the percentage on-time and the number of occurrences in an 8-hour period either for night time or the rest of the day.

Having done this, one tries to get the noise switched off so that the ambient or background noise level can be measured. This will not be steady either, and it is necessary to watch the needle of the meter and record the typical low value, ignoring occasional jumps of the meter. Anyone who has tried to get a gas works shut down so that he can see what the background noise level is without it will know that it is sometimes necessary to by-pass this stage in the procedure.

One's sojourn on the site is then over, and assessment of the noise is then carried out on paper. A figure called the corrected criterion is calculated, starting with 50 dBA as the basic criterion; 5 dBA is added to it if the factory has been established a few years but is not typical of its area, and 10 dBA is added in the case of old-established factories which are completely in character with their surroundings. The foundry referred to earlier would come into this category. One of course includes any industrial premises or fixed installations under the heading of 'factories'. Nothing is added to the basic criterion in the case of new factories, or of existing factories being altered in such a way as is liable to increase the noise level, or where a new process is being installed.

More corrections are made then to take into account the 'noise-climate' of the neighbourhood. 5 dBA is subtracted in the case of rural residential districts, nothing in the case of suburban districts with little road traffic, 5 dBA is added for urban residential areas, with further categories up to 20 dBA added for predominantly industrial areas with few dwellings. On top of this 5 dBA is added if the noise occurs only during working hours, and is subtracted if it occurs at night, and finally, 5 dBA is added if the noise occurs only in winter, presumably because people tend to keep their windows shut more and do not sit out in the garden if they can help it.

Now one gets down to what the Standard is all about: rating the noise. Firstly one compares the corrected background noise against the corrected criterion. This gives you an idea of whether the noise-climate is typical of the circumstances. If the corrected background noise level exceeds the corrected criterion, that is, if the area is unusually noisy, provided the excess is within 10 dBA, complaints may be expected if either of our two corrected levels, now designated L'_1 and L'_2, exceed the corrected criterion by 10 dBA or more. If the reverse is the case and the area is unusually quiet, complaints may be expected if either L'_1 or L'_2 or both exceed the background noise level by 10 dBA or more. If, however, you have not got a figure for the background noise, then you rely entirely on the corrected criterion, and if either L'_1 or L'_2 exceed it by 10 dBA or more then trouble may be brewing. If it is the other way round, then complaints will be extremely unlikely.

In all cases, excesses of 5 dBA are considered of marginal significance, but of course if one tolerates an excess 5 dBA on the noise from every factory there will be a liability for a creeping background noise to develop. Also, in places where although you cannot measure the background level it is obviously abnormally high or low, it would not be very meaningful to rely on the corrected criterion only.

If you do find you have a problem, the dBA measurement is no longer any use, and in order to locate the frequencies at which the noise reduction is required one has to go on and analyse the noise into frequency bands, using octave, half-octave, one-third-octave or even narrower-band filters. The British Standard gives a chart which has as its principal merit the fact that it is simple to draw. It constitutes a very much simplified set of equal-loudness curves on which one plots the results of an octave-band analysis, and can get an idea of the band which contains the most troublesome components.

The figures in the Standard are undoubtedly on the high side, but it still has considerable uses in predicting the likelihood of complaints, or the setting of criteria. It is also seriously abused. The entire procedure is designed purely to indicate whether a noise is likely to be acceptable to typical groups of people. It does not lay down permissible levels. Unfortunately one comes across more and more cases where a noise has been rated according to the Standard and found not to be excessive, but where the raters, whoever they may be, then say to the poor noise victims: 'According to British Standard 4142: 1967 you are not supposed to be complaining!' Of course this is nonsense; the Standard is a guide to those people whose job it is to specify criteria and to deal with noise to help them to predict what will and will not be acceptable. It is not in any shape or form a law which says whether or not a householder has a right to complain, even if it may occasionally be referred to in court proceedings.

The Standard is by no means the only guide to the assessment of noises, and over the years a fair number of systems have been evolved. In fact, it would be possible to write a whole book on the subject of acoustic units and rating systems. We have come across dBA and made reference to PNdB, phons, sones and so on. There are also batches of curves such as noise-criterion curves, noise-rating curves and several others. However, it is becoming increasingly evident that the dBA wins as the most useful system. The PNdB is a close second, and both have the immediate advantage that they can be read directly from a meter. Almost all meters incorporate an 'A' weighting network, and meters are just beginning to have a 'D' network, enabling PNdB to be read directly. The PNdB unit is liable to be modified in the near future, but broadly speaking it is like a dBA but with extra sensitivity to sound in the region of 4 KHz.

On the other hand, there is a lot to be said for the I.S.O. proposed noise-rating curves, which are not covered in detail by this book. They approximate to simple equal-loudness curves. A noise is assigned an NR number when its octave-band analysis is plotted on a graph of NR curves and it exceeds one of the rating curves in one or more octave bands. NR curves are much easier to use than dBA because there are almost infinite permutations of octave-band levels which go to make up the same level in dBA. However, for our present purposes, dBA are more than adequate.

The final report of the Wilson Committee on Noise makes some tentative recommendations on the subject of noise levels in dBA inside living-rooms and bedrooms, as opposed to the British Standard criterion for levels outside houses. The Wilson Report very tentatively estimates that the following levels should not be exceeded for more than ten per cent of the time:

Situation	Day	Night
Country areas	40 dBA	30 dBA
Suburban areas, away from main traffic routes	45 dBA	35 dBA
Busy urban areas	50 dBA	35 dBA

This estimate was made in the realization that to meet these levels a 5 dBA or more reduction in noise levels would normally be necessary.

Of course traffic and factories are not the only cause of noise disturbance in the home; the neighbours can create quite a racket if they feel like it, and sometimes even if they do not they can inadvertently annoy their neighbours simply by flushing a lavatory or even walking about. The Building Research Station at Watford carried out a number of surveys as a result of which we have a grading system for sound insulation for walls and floors between dwellings. There are three grades, a House Party Wall Grade, Grade I and Grade II. Grade II is about 5 dB worse than Grade I and the House Party Wall Grade is Grade I with an improvement at low frequency, 4 dB at 100 Hz. A 230mm brick party wall between houses normally gives sound insulation up to the House Party Wall Grade.

If you remember that according to the mass law a 5 dB drop in insulation is the result of halving the surface density of the wall, then you can see that a Grade II wall is a very much lighter affair. Much the same applies to floors: a 120mm concrete floor with a resilient quilt and floating floor on top of it will tend to satisfy Grade I, and a solid 200mm concrete slab would do about the same. The application of the House Party Wall Grade speaks for itself. Walls separating flats and maisonettes or bedrooms and living-rooms from common halls or corridors should make Grade I, and this will result in noise only being a minor nuisance. If only Grade II can be obtained noise will be much more of a nuisance, but probably more than half of the tenants will not be seriously disturbed.

One is thus fairly well guided when it comes to noise in the home, but what about noise at work? In chapter 5 we talked about hearing damage, but obviously it is not enough simply to ensure that one's staff are not in danger, even if this is our prime duty. The figure of 90 dBA is appearing more and more as a maximum safe level for the general noise climate in a place of work. Chapter 5 showed that the assessment of a hearing damage risk is really a very complex business, and in certain cases, where there is a decent percentage of off-time for instance, levels higher than 90 dBA would be safe. Nevertheless, this figure may well appear in future legislation. There is, once again, the danger that two different noises, one more dangerous than the other, would both amount to the same level in dBA. This is where the need for frequency analysis and noise-rating curves comes in, but as any sort of frequency analyser costs at least £200 it is much better to use units which can be read from an instrument costing a quarter of the price. A very quick guide to a hearing damage risk is that if staff who do not normally work in the area concerned but make periodic visits to it come out with a temporary threshold shift (chapter 5), the situation should be looked at very closely, if necessary by a specialist.

The next on the list of priorities is efficiency and general contentedness. Once again, as far as factory personnel are concerned, 90 dBA appears to be the level above which noise impairs efficiency to an extent which is easily demonstrated. It is probably not a coincidence that this is the same level as the hearing damage criterion. This assumes that the work being done is largely the operation or minding of machines. Of course, as the need for concentration increases, so does the need for quiet; I could not write this book in a noise climate of 90 dBA, and am in fact fortunate enough at this moment to be looking at a sound-level meter registering 32 dBA.

If one starts at the top of the scale, 90 dBA can be reserved for heavy industry or seriously noisy processes where the cost of noise control is so high that nothing short of ensuring safety is justified; 80–85 dBA is just about acceptable in light-engineering works, 75 dBA on the high side for intricate work such as the assembly of electronic components. Many general offices have levels in the 60–65 dBA range, but these are usually complained about; 55 dBA is more acceptable. Smaller offices are best kept at 45–50 dBA, and private executive offices at 35–40 dBA.

The noise which goes to make up these levels, particularly at the lower end of the scale, should be devoid of particularly irritating characteristics such as impulsiveness or tonal character. If these attributes do exist, the offending noise ought to be kept 5 dBA lower than usual. If, as happens with things like type-writers, it is not possible to reduce the impulsive noise, it is better to try to raise the background level to act as a mask, thus removing some of the nuisance value of the noise. A general office with an average background noise level of 50 dBA but with impulsive noises from typewriters and people opening filing cabinets or dropping things on the floor superimposed will be less satis-factory than a similar office with an ambient of 55–60 dBA, say due to ventilation noise, which is devoid of irritating character-istics.

The ideal office situation is one where the reverberation level is very low, because of an acoustic ceiling and some sort of carpeting, and where the resulting loss of general background hubbub is made up for by the introduction of a mask from the ventilation system. In fact it has even been found necessary in some offices to generate the masking noise with tape recorders, or electronic sound generators. This sort of thing is particularly important in large landscaped offices where up to 200 people may work in one room.

So far this chapter has contained purely guidance and advice. Are there no laws about noise? Surprisingly enough, existing laws are able to cover a very wide range of cases. The only mention, at the time of writing, of actual limits in decibels is in the Motor Vehicles (Construction and Use) Regulations 1969. These limit, among other things, the noise from motor cars, measured under stringently specified conditions, to a level of 85 dBA. Unfortunately, complying with the test method as laid down in B.S. 3425: 1966 is no easy matter, if only because you have to find a space of at least 50m radius of which the central area of 10m radius consists of concrete, asphalt or similar hard material free of any soft covering such as snow, long grass, loose soil or ashes. There have to be no substantial obstructions within 25m of the vehicle, and '. . . the sum of the angles subtended at the position of the test vehicle by surrounding buildings within 50m radius shall not exceed 90°. . .' This is why you do not often see noise traps operated by the police in the same way as radar traps.

However, there is a much greater section of the law which relates to noise than this. Much can be written on the law relating to nuisance, which covers an amazing range of things from keeping a brothel to exposing a person with an infectious disease in the streets, or simply obstructing the highway (whether or not in the pursuance of the previous two activities).

There are three types of nuisances: public nuisances, statutory nuisances and private nuisances, but in fact the word 'nuisance' is the only thing they have in common. There is no clear-cut definition of each, but in practice they can be defined as follows:

(1) Public nuisance: Inconveniencing or endangering the public.

(2) Statutory nuisance: Nuisances designated as such by Statute.

(3) Private nuisance: Interference with another person's enjoyment of his land.

Public and statutory nuisances are crimes; private nuisance is a tort and the aggrieved person must take action by way of the civil courts. The Attorney-General is the only person who can take action over a public nuisance, and although he may indict the offender he normally applies to the High Court for an injunction.

All three types of nuisance cover noise. Because public nuisance is a crime and private nuisance is a tort, it is obviously more important to know the difference between the two.

The then Lord Justice Denning said: 'The classic statement of the difference is that a public nuisance affects Her Majesty's subjects generally, whereas a private nuisance only affects particular individuals. But this does not help much. The question "When do a number of individuals become Her Majesty's subjects generally?" is as difficult to answer as the question "When does a group of people become a crowd?" ... So here I decline to answer the question. ... I prefer to look at the reason of the thing and to say that a public nuisance is a nuisance which is so widespread in its range or so indiscriminate in its effect that it would not be reasonable to expect one person to take proceedings on his own responsibility to put a stop to it, but that it should be taken on the responsibility of the community at large.'

In order to take action for nuisance there must be damage, even if this includes interference with the beneficial use of the plaintiff's premises. When the nuisance is a noise, the basis would normally be one of interference with personal comfort. This has got to be substantial.

As far as statutory nuisance is concerned, the Noise Abatement Act 1960 brought noise within the scope of the Public Health Act 1936. This gives a local authority the right to serve an abatement notice on the perpetrator if they are satisfied that a noise nuisance exists. The abatement notice must specify any works required to be carried out and if the nuisance is not stopped the authority may make a complaint in a magistrates' court. Alternatively any three people may take the latter action and institute proceedings. The court may then make a nuisance order requiring that the abatement notice be complied with, and throw in a £5 fine, and if this order is not complied with the local authority may itself carry out whatever works are necessary to abate the noise.

What about defence? If you commit a nuisance and find yourself in court, what can you do? The main course of action is to try to show that the act complained of was not a nuisance at all, or is another sort of nuisance. You can, in a case of public nuisance, use the defence that it was not a public but a private nuisance. In a case of statutory nuisance, you can use the defence that you have taken the best practical means for preventing and for counteracting the effect of the noise. There are one or two other defences available, but they are not very relevant when the nuisance is noise. One of them is that there exists a state of affairs which could not have been prevented by the exercise of all due care and attention.

However, where a statute authorizes the carrying-on of a particular act which will inevitably involve the committing of what would normally be a nuisance, then, provided that every reasonable precaution has been taken by those carrying out the act, there is no way in which they can be taken to court for nuisance of any kind. Those who carry on these activities are known as statutory operators, and include the operators of aircraft and railway trains, and the gas, electricity and water authorities. You will thus not get very far if you try to instigate an action for public nuisance against British Rail, for instance.

There are two main remedies for nuisance, one of which strangely enough does not involve the courts. If you are the victim of a nuisance you can take the law into your own hands and carry out your right of abatement. If the man next door commits a nuisance, say by having a noisy fan in a wall of the house, you may enter his land and jam it if the noise is making life intolerable. However, this action would be most inadvisable and you would almost certainly lose any subsequent right of action in the courts.

By far the most common remedy for noise nuisance is an injunction, which is a judgement or order of a court restraining the committing or continuing of some wrongful act or omission. Failure to comply with it will lead to imprisonment. An injunction will not be granted if the damage complained of is trivial or if it would be oppressive to grant it. It is often accompanied by an award of damages, and where an application for an injunction is accompanied by a claim for damages it can then be granted by a county court judge – there is no need for an application to the High Court.

Common Law nuisance is by no means the only way in which cases concerning noise may be brought to the courts. However, for some reason actions for negligence or under the provisions of the Factories Act 1961 are exceedingly rare. The Factories Act does not mention noise, but it raises considerable possibilities on this count. Section 54 (1) states:

'If on complaint by an inspector a magistrates' court is satisfied either –

'(*a*) that any part of the ways, works, machinery, or plant used in a factory is in such condition or is so constructed or is so placed that it cannot be used without risk of bodily injury; or

'(*b*) that any process or work is carried on or anything is or has been done in any factory in such a manner as to cause risk of bodily injury; the court shall, as the case may require, by order –

'(i) prohibit the use of that part of the ways, works, machinery or plant, or if it is capable of repair or alteration, prohibit its use until it is duly repaired or altered; or

'(ii) require the occupier to take such steps as may be specified in the order for remedying the damage complained of.'

No one would dispute that damage to hearing is bodily injury, but the Factory Inspectorate have never, to date, made use of the relevant sections of the Act. Fortunately, Factory Inspectors are able to obtain considerable cooperation from industrial concerns, as are Public Health Inspectors in the field of nuisance. However, the lack of any precedent has made the Factory Inspectorate afraid of what may happen if they attempt to obtain an order and fail. However, there is likely to be a new Factories Act which will cover noise specifically and quantitatively, so that this problem will be resolved.

As far as actions for negligence are concerned, to date there has only been one, which was unsuccessful. There are one or two more in the pipeline. This one case, though, is very important, and it is worth recounting the details. Broadly speaking, in a Common Law action for damages as a result of noise-induced deafness, the plaintiff has the burden of proving that his deafness resulted from the noise exposure in question, and that his employers had been negligent with respect to the noise exposure. This means that he must prove that his employers should have recognized that there was a hazard and should have taken appropriate preventive action in the light of the particular medical and technical knowledge available at the time of injury. The only Common Law action that has ever been brought for noise-induced hearing loss until 1969 is that of Down *v.* Dudley, Coles, Long Ltd, which was heard before Mr Justice Browne at Devon Assizes on 27–31 January 1969.

The unfortunate Mr Down had been employed for two weeks in May 1966 on a construction site at Exeter University. His work consisted of the fixing of sheets of expanded metal to concrete window lintels, and this was achieved using rivets fired from a cartridge-assisted hammer. He fired about 130 rounds of ammunition a day, and because of the level of the work he had to hold the hammer close to his right ear. About half the shots were fired up in the corner formed by the wall and ceiling, with the result that the impulsive noise from the hammer reached peak sound pressure levels of over 160 dB.

Unfortunately Mr Down's hearing was not very good to start with, but his right ear was the better of the two. He soon, understandably, found the noise from the hammer a bit much, and asked his foreman for some cotton wool with which to plug his ears. This he was given, as neither he nor his foreman realized that cotton wool would not really have any effect at all. He then continued with the work and towards the end began to experience extreme dizziness, which lasted for the best part of a week; he also totally and permanently lost the hearing in this previously better right ear.

The action for negligence was brought on the basis that the employer was negligent in failing to appreciate that Mr Down's work created a risk of hearing damage, in failing to take any precautions to avert the risk (not appreciating that cotton wool would provide no protection), and not providing suitable ear plugs. In addition it was claimed that the foreman was negligent in not either removing him from the work when he was obviously being affected by the noise (otherwise he would not have asked for cotton wool), or giving him proper ear plugs. In law the employer would be liable for the foreman's negligence.

The prime question was, as Mr Justice Browne said in giving judgement, 'Ought the defendants to have foreseen that the use of this gun described by the plaintiff might have caused injury to his hearing?' No one disputed that the noise was a serious hazard to hearing; expert witnesses appeared for both sides, Surgeon

Commander R. R. A. Coles for the plaintiff and Mr H. D. Fairman for the defence, and they were in agreement that the probable cause of Mr Down's deafness was the noise from the gun. Unfortunately, though, no evidence was offered to suggest that this noise was generally recognized as hazardous. The pamphlet 'Noise and the Worker' published originally in 1963 by the Ministry of Labour says 'Ear plugs are designed to occlude the ear canal and must be correctly fitted. . . . They may be made of rubber, neoprene, plastic or cotton impregnated with wax – not dry cotton wool, which affords little or no protection', but it was considered that the pamphlet was too vague in the paragraph, 'Have you a Noise Problem?'. The relevant British Standard mentions noise hazards but was not published until November 1966, and one or two scientific papers on the subject were not available at the time.

The judge found that the defendants were not negligent in failing to appreciate at that time that the gun noise was a hazard and therefore the question of their negligence in failing to appreciate the uselessness of cotton wool was irrelevant. There remained the question whether or not the foreman was negligent in not taking Mr Down off the job or giving him proper ear plugs when he asked for cotton wool. In giving evidence the plaintiff recounted his words as 'I think I will have some cotton wool and stuff my ears up. I think that will be a protection. I said I wanted protection against the noise of the gun.' The judge decided that the question was whether Mr Down had meant 'I want protection against discomfort and inconvenience' or 'I want protection against risk of injury'. He was satisfied that the foreman ought to have construed the words as meaning the latter and therefore ought to have done something about it, but concluded his judgement by saying that 'it may not be unreasonable for a foreman to whom that was said to fail to understand the plantiff may be sustaining some injury to his hearing'. The foreman was unaware of Mr Down's already defective hearing, which might have alerted him to the possibility of some risk, and the judge with great regret gave judgement for the defendants.

It is most unfortunate that this had to be the outcome of the first Common Law negligence claim for noise-induced hearing loss; there are more cases to come, and meanwhile it is every acoustician's duty to remove the cause of the failure of Mr Down's case, and indeed the cause of his deafness: ignorance.

Is, then, the law adequate to deal with noise? Common Law of nuisance is quite adequate to deal with questions of annoyance of the public and its individual members. Common Law of negligence can be made totally adequate, not by Acts of Parliament, but by education of employers, so that they cannot escape with a plea of ignorance of a hazard. The provisions of the existing Factories Act and Offices, Shops and Railway Premises Act could be invoked by the Minister to enable wide-ranging control over noise in workplaces. It is difficult to see how the Motor Vehicles (Construction and Use) Regulations 1969 could be improved upon; the trouble is that the police have neither the time nor the inclination to enforce it. Most mention of aircraft in Acts concerns the permitting of noise rather than the control of it, but there is now a proposal for a noise certification scheme for aero-engines which will force some modification to many existing engines.

As far as the future is concerned, we shall certainly see new legislation, not least of which may be a new Factories Act which will follow the precedent set by the Motor Vehicles Regulations 1969 and incorporate statutory limits in dBA.

Generally speaking, the law is already capable of dealing with a surprisingly wide range of noise cases. The public and the Government must make use of it.

12 How quiet can it be? – how to design quiet machines

How do you design a quiet machine? What extra something do you have to put into it to cut down the noise?

We now know about the three stages in the making of a noise: the initial disturbance, amplification or modification of it, and radiation. Unless your philosophy is such that you take aspirin to relieve the discomfort of sitting on a drawing-pin, your mind will immediately turn to thoughts of going to work on the initial disturbance. This is of course the best approach, but how often can it be done? Take the diesel engine, for instance. The initial disturbance, the pressure rise in the combustion chambers, is the very source of power in the machine. In the turbo-jet the burning of fuel is the sole means of providing the required acceleration of gases at the rear. In the pneumatic drill, repeated blows on the bit are what the whole machine has been designed to bring about. The prospects certainly do not look very promising, but in reality there is more room for improvement than you might think.

In chapter 6 we looked at graphs of cylinder pressure in a diesel, and saw how these could be analysed into pure-tone components of suitable phases and amplitudes by means of Fourier analysis. Figure 24 showed that it was the amplitude of the higher-frequency components which affected the level of noise cause by excitation of resonances in the engine structure. If the Fourier analysis or frequency spectrum can be made to fall in magnitude as sharply as possible with increasing frequency, great benefits in reduction of noise, up to 15 decibels, can be reaped.

Look at figure 48. This shows three cylinder pressure graphs for a particular diesel engine running at 1,000 r.p.m. Curve 'a' is for a condition when the engine is driven by an external means and no fuel is being injected; air is merely being compressed as

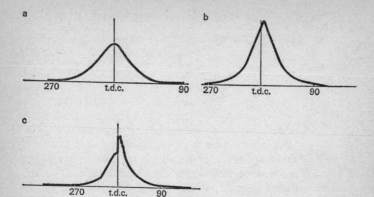

48. Cylinder pressure oscillograms (diesel). (a) Motored. (b) Smoothest possible curve with engine running. (c) Most abrupt curve with engine running. All at 1,000 r.p.m.

the engine is motored. This is the smoothest cylinder pressure curve that could ever be achieved, and figure 49 shows that the result is an extremely abrupt fall in the level of the Fourier components above 100 Hz. Unfortunately a diesel engine under these conditions would not be much use. Curve 'b' shows the pressure graph for the same engine, running this time, producing about the smoothest curve possible in a diesel engine. Notice the slightly sharper peak, and the steep flat section around top-dead-centre. This is the moment, immediately following injection of the fuel, when combustion occurs. In this case it takes place fairly progressively over a period of some 20 degrees. Figure 49 shows the frequency spectrum for this condition. Because of the less smooth shape of the pressure curve, the higher components in the frequency analysis have a greater intensity by over 20 dB at high frequency compared with curve 'a'. Apart from the 'kink' in the curve at 1,600 Hz, curve 'b' compares well with that of a petrol engine with the same pressure rise. The 'kink' is due to resonance of the gas in the combustion chamber, which 'sloshes' to and fro.

Now look at the horror in this trio. Curve 'c', where the combustion of the fuel takes place almost instantaneously at t.d.c.

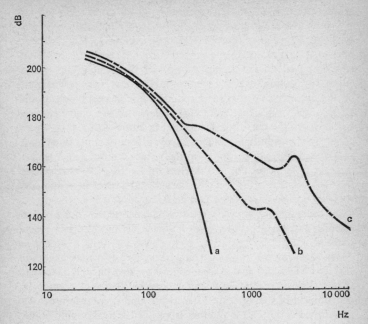

49. Frequency spectra of the curves in figure 48

There is a sharp almost vertical rise in pressure of over 28×10^5 N/m², followed by marked oscillations superimposed on the main curve. Not surprisingly, this curve contains Fourier components of great magnitude at high frequency, with a large 'kink' at 2,600 Hz. The frequency spectrum is not much better than that of a saw-tooth electrical waveform which would be the ultimate in noisiness that any fiendish engine designer could aspire to!

So the picture turns out to be much less hopeful as far as the initial disturbance is concerned than it first appeared. However, how do you smooth out the pressure rise? The first variable is timing. In curve 'b' injection occurred earlier than in curve 'c', and, although this was not the only reason for the difference between the two, it is easy to see why it is significant. If injection takes place only fractionally before top-dead-centre there will be a higher pressure in the cylinder to start with. This will not only mean a higher temperature and faster ignition, but also the pressure boost caused by the sudden burning of fuel and air will be superimposed on the highest section of the basic compression curve. If injection takes place earlier, the 'boost' is superimposed on the compression curve to produce a less jagged shape. Look again at curve 'c'. Just before the violent rise at t.d.c. there is a flattening of the curve similar to that in curve 'a'. In fact if you cover up the right-hand half of curve 'c' it looks just like curve 'a'. The pressure boost comes at the very worst moment because the rate of compression had levelled off as the piston arrived at the top of its stroke. If you take the same pressure boost and add it to an earlier part of the compression curve where there is a steep pressure rise anyway, the two blend together much more happily and the level of higher-frequency components is much reduced.

Timing is not the only governing factor by any means. Another is the way in which fuel is sprayed into the combustion chamber. If it is sprayed on the walls of the chamber, there is a less even distribution of fuel and air. This means that combustion occurs at a slower rate than if the fuel is sprayed into the middle and almost instantaneously creates the right fuel/air mixture. Combustion becomes a more progressive business and the slope of the pressure rise curve is not so great. Subjectively this results in the almost complete elimination of the 'knock' in the diesel engine and is similar in principle to the prevention of pinking in a high-compression petrol engine. Here the use of high-octane petrol controls the combustion of the fuel and prevents premature spontaneous ignition in the high temperature caused by the high compression.

In theory, the smoother the cylinder pressure curve, the more efficient should be the engine. It is a noise engine after all, and it is the conversion of the fundamental frequency of the cylinder pressure oscillations into mechanical rotation via the pistons and crankshaft which provides the power. Energy in the form of higher harmonics is useless, because the piston is oscillating in the cylinder only at the fundamental frequency. Whenever the pressure curve is smoothed out, it results in proportionately more of the energy being concentrated in the lower harmonics which can be transmitted to the crankshaft.

What about other initial disturbances, in the turbo-jet for instance? There is a difference here: although the combustion of fuel is essential to the operation of the jet, pressure fluctuation is not. Constant pressure in the cylinder of a diesel engine would be useless, but constant pressure in the combustion chamber of a jet would be ideal. A jet does not rely on pressure fluctuations of any sort for its power. It is merely the limitations of modern engineering science which cause them.

The pressure variations which cause the greatest trouble do not take place in the combustion chamber, but after the gases are ejected from the nozzle and mix turbulently with the air outside. A vast amount of research has been and is being conducted into means of bringing about smoother mixing downstream of the nozzle. This has led to many designs of nozzle, but the overriding factor is of course the effect any modification has on the thrust of the engine. An early development involved the use of a deeply corrugated nozzle (figure 50) which facilitates the mixing of gas

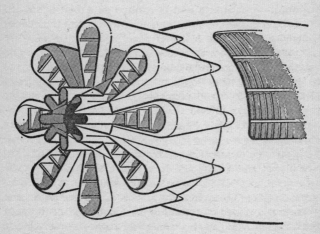

50. An eight-lobe jet-engine silencer

and air by drawing the air deeper into the jet of gas immediately downstream of the nozzle. This can reduce noise in the mid-frequency range by 6 or 7 dB. It has also been found that reducing nozzle diameter shifts the noise higher up the spectrum, and as air absorption of noise is greater at high frequency this can help. On the other hand human sensitivity to noise is greatest at around 4,000 Hz and as the greatest aircraft noise problem occurs when the jets are near the ground this method of approach would on balance make matters worse.

By far the greatest stride forward in the reduction of jet noise has been made with the introduction of the turbo-fan engine, such as the Rolls-Royce RB 211 powering the Lockheed TriStar airliner. This engine has what is known as a high by-pass ratio. The first stage of the compressor is greatly enlarged to be more of a multi-bladed fan than a compressor stage. This is surrounded by a large cowl. Then follows the rest of the compressor, which is smaller in diameter, as is the body of the engine. A large proportion of the air handled by the first stage is spilled out over the body of the engine through the back of the large cowl. Only a small amount of air enters the engine. The result is that although a large mass of air is accelerated by the engine, only part of it goes through the combustion chambers and is emitted as a high-speed jet. Most of the air passes at ambient temperature along the outside of the engine and creates a 'buffer zone' between the turbulent jet and the atmosphere when it reaches the rear of the nozzle. For a start, the air which has been by-passed is not so turbulent as the jet, therefore reducing noise, but also the mixing of the gases of the jet with the cold and atmospheric air is much smoother because of the intermediate layer of by-passed air.

Things would certainly be simpler if jet noise were the only source of noise from a turbo-jet (or turbo-fan) engine, but it is not. There is also the disturbance created by the interaction of compressor blades, and, to a much lesser extent, turbine blades. The latest generation of Rolls-Royce engines embodies a major in-novation on this score, because the normal inlet guide vanes have been entirely omitted, eliminating an important disturbance caused by each blade suddenly meeting a drop in air velocity every time it passes behind a stationary vane. Noise from rotor/

stator interaction, where it does occur, has been reduced by using wider spacing between rotors and stators, and by incorporating ideal ratios between the total number of blades of each.

These engines also embody three separate concentric hollow drive shafts, so that fan speed can be reduced as it is driven by a separate shaft from the compressor. In addition there is a device for reducing nozzle area and shape, changing from circular to rectangular – by changing the radiation pattern so that most of the noise goes out sideways, a reduction of 3 dB is obtained on the ground. Noise is still further reduced in this engine by means of a large amount of sound-absorbent lining which is built in to the various ducts. As we shall see in the next chapter, lining a duct with sound-absorbent material can be very effective in attenuating noise travelling along the duct. Some 20 sq. metres of sound-absorbent lining is incorporated in the engine (figure 51),

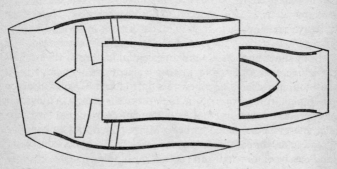

51. RB 211 noise-absorbent linings (shown by heavy lines)

in the inlet, fan and exhaust ducts. Porous plastic fibrous materials are used in the low-temperature duct, metallic fibre in hot areas. As chapter 8 showed, the frequency range of fibrous absorbers is lowered by spacing them away from the backing. The lining to the fan duct, which handles lower-frequency noise than the exhaust duct, is therefore spaced away from the walls, but in the exhaust duct there is no need for this on account of the high frequencies of the turbine noise. Of course the really low-frequency jet noise is generated downstream of the engine, and no amount of sound-absorbent lining would affect it. However, because of the high by-pass ratio, jet noise in the RB 211 takes second place to fan, compressor and turbine noise.

What about initial disturbances in other machines? In a pneumatic drill there is nothing effective you can do to the basic mechanism – compressed air acting periodically on a piston – which does not radically reduce the efficiency of the drill. Although it has similarities with a diesel engine in that one can produce cylinder pressure diagrams (which would be far more jagged), the means by which these pressure fluctuations are turned into noise are very different and any method of smoothing out the curve would not be worth while.

There are, however, other machines to which a great deal can be done to reduce the disturbances. In a punch press, for instance, different parts of the same die can be varied in height so that they do not all strike the workpiece simultaneously. The sharpness of the impact is greatly reduced. Of course wherever steady progressive pressure, as in hydraulic machines, can be used in place of a sudden impact, the noise problem is almost eliminated.

Sometimes impacts are not an essential feature of the machine in which they occur. No loss of efficiency is then involved if cushioning washers and pads are introduced. They will accept an impact from a component and absorb the energy in it by being compressed, when internal damping causes frictional losses and the energy is dissipated as heat. Many punch presses contain a great many impact noise sources which are quite unnecessary and can be eliminated with rubber* cushioning pads.

However, for this approach to be any use except at high frequency, it is essential that the rubber pads or washers are thick and soft enough to be given a chance. Obviously if no sizeable compression occurs there will not be much dissipation of energy and therefore it is sometimes necessary to use up to 20–30mm of rubber. It is essential to realize that rubber is virtually incompressible. This may sound like saying that chalk is cheese, but it is true. Rubber can only be deformed in a large degree when subjected to shear stress, not direct compression. If you place a small

* The term 'rubber' is here used to cover a wide range of polymers as well as natural rubber.

block of rubber on the floor and stand on it, it will be squashed, but only because it bulges out at the sides. It is yielding to shear forces at right angles to the direction of the applied force. If you were to put a steel band round it, you would soon see that it could barely be compressed because the sides could not bulge.

This is a very important fact about rubber, and it will crop up several times. At this stage we are talking about absorbing impacts, and the first important thing is that a large piece of dense rubber in compression will be elastic only in proportion to the area of the sides which are able to bulge. If they cannot bulge or if the area is small, as in a large sheet of rubber, you can regard the rubber as solid. For this reason, foam rubber is often more useful in impact absorption because the air pockets allow the rubber to deform in all directions. For light loads, much greater deflection can be achieved, but it is just as important not to use a foam which is so weak that it becomes solid under load. Sadly, though, there are a great many cases where sharp impacts are essential. A typewriter, for instance, would be much quieter if it had a foam rubber platen. It would also produce a very poor impression.

The reduction of aerodynamic disturbances, as we have seen with the jet engine, can be very difficult. 'Streamlining' of solid obstacles in relative motion with air is the obvious first course, and good aerodynamic design of fan blades can at least eliminate unnecessary vortex formation. It is particularly important that the angle of attack of the blades is not too great. An effective method of cutting aerodynamic disturbance when there is a rotor/stator interaction is to increase the clearance between the two, whether they be fan blades and struts, or the rotor and stator of an electric motor. Once again, though, one can be thwarted by other considerations, and separation of rotors and stators can upset the efficiency of machines. In some centrifugal fans, there is an improvement in the level of the blade-pass note if the clearance between the rotor and the cut-off point of the scroll is increased. It is also important to avoid turbulence in the air upstream of a fan. This is because vortices can be large enough to have an effect on the lift of a blade; variations of lift mean pressure fluctuations, and noise.

It is certainly nearly always possible to improve matters by careful arrangement of the number of components of the rotor and stator. A four-bladed fan running close to four guide vanes would produce a much worse blade note than it would if there were three or five guide vanes. As a general rule, the number of components of either rotor or stator must not bear a simple ratio to that of the other. This will mean that not only will the fundamental frequencies caused by the two be different, but so will the first few harmonics. The benefit is twofold: the intensity of the pure-tone components of the noise will be less, and the nuisance value of the noise will be reduced because the spreading of the tones and harmonics will make them less prominent.

When it comes to mechanical vibration as the initial disturbance, there is often much to be gained by tightening up manufacturing tolerances. Ball bearings are noisy because of irregularities in the shape of the balls and the races, which will one day be eliminated, even if it means casting the balls in a weightless state in an orbiting factory! High-speed rotation noise can be reduced by better balancing. Wherever a fluctuating force is present, the principles we looked at earlier in the diesel apply. If the fluctuation can be pruned of any sharp features, the higher Fourier components are reduced in intensity. If the regularity of the fluctuations can be broken up, the sound energy can be diffused over the spectrum and not concentrated at only a few frequencies.

Next time you examine the tyres on your car, or before they wear out, look at the pattern of the tread, particularly at the edges. You will see that the transverse cuts between the sections of rubber are by no means regularly spaced, but that the distance alternates progressively between narrow and wide. This is done to prevent the tyre from 'whistling', instead of which it produces more of a random hiss.

A section of the M10 motorway was given an experimental surface which was basically concrete with transverse slots to drain water away. Unfortunately the slots gave rise to loud whining noises when driven over, and anxious drivers were to be seen on the hard shoulder peering at their back axles. Another section was tried with simple alternate spacing distances between slots, but as would be expected this still produced pure tones. Only if either a completely random or a carefully calculated progressive alternation between narrow and wide spaces is used can this problem be overcome. The latter method has the effect of 'warbling' the note fast enough for it to sound more like broad-band noise, while both methods distribute the sound energy over a wider frequency range. The problem is not, as was thought, one of tyre resonance.

This introduces the interesting possibility of providing talking road signs by serrating the surface of the road in a manner similar to the undulations in the groove of a gramophone record. The voice from the road would unfortunately vary from Paul Robeson at slow speeds to Donald Duck at high speeds. There could be permanent signs cast into concrete, and variable signs incorporating slats of variable height which could be hydraulically raised proud of the road surface. So long as this system could never become an advertising medium it would have many advantages, for instance in the control of 'fog-hogs' on motorways.

This type of initial disturbance is found in many machines. If the blades of a circular saw were not strictly regular, carefully calculated spacing of them could spread the frequency content of disturbance, both aerodynamic, and as vibration of the blade. For this reason many engine cooling fans on motor cars have a strange-looking irregularity in the blade positioning. The same can be done to the teeth of milling machine bits, routers and drills.

However, if all these initial disturbances could in reality be detached from their amplifiers and loudspeakers life would be much quieter. If we cannot do much to reduce them, all is not lost, because in many machines dealing with the two other links in the noise-production chain can be quite effective enough.

Let us go back for the last time to the diesel engine and see how the amplification and radiation mechanisms can be tackled to reduce noise emission. We know that the full brunt of the noise level in the combustion chambers does not reach the outside world and that this is because of the difference in impedance between the air in the combustion chambers and the crankcase and block. We also know that the most important factor which reduces the impedance of the crankcase and block is resonance. A major step forward will be achieved if some form of damping can be introduced into the construction of the crankcase and block to reduce resonance.

A great deal of research has been done on this subject by C.A.V. Limited, makers of fuel injection equipment for diesel engines, who in order to prove that injector noise was not the main culprit in diesel noise, set about constructing a quiet diesel engine. Their attempts were extremely successful and they constructed two types of engine which produced a noise level that on a subjective scale was roughly half the loudness of comparable conventional engines. They tackled the problem in two ways. The first was to construct an engine with highly damped walls, and the second to use materials and conformation for the crankcase and cylinder block which would greatly increase its stiffness and push its resonant frequencies up by about a factor of five.

Two engines were constructed on the first principle which in most respects were similar. The basic framework of a three-bearing four-cylinder engine was welded up from steel plates, to house the wet liners of the cylinders and the bearings for the crank and camshaft. These parts, as well as the cylinder head, were standard. This skeleton-type crankcase and cylinder block was thus able to withstand all the stresses required of it, and the only thing it would not do was to hold the oil and water in. Because the total external surface area of the skeleton was very small, it was not able to radiate much noise.

The problem was then simply to find materials with which to form the side walls of the structure which would not amplify the vibrations in the skeleton through resonance. Unfortunately any walls would increase the noise to a certain extent by providing the extra surface area lacking in the skeleton.

Sets of interchangeable panels were made up of materials from steel, lead and plastic to a steel/rubber/steel sandwich. The lead and plastic panels had high internal damping and were extremely effective, but the last method, using laminated steel and rubber, was finally adopted. These panels were fixed to the edges of the skeleton by a number of small screws and were found to emit up to 15 dB less noise than the cast-iron sides of a conventional crankcase. In addition, the sump was isolated from the main structure by means of a rubber strip, and other large metal covers such as the timing cover were highly damped, being constructed from the same 'sandwich' of steel and rubber.

One of the places in a diesel engine where the highest levels of vibration occur is in the crankshaft itself. This is as one would expect, because everything that happens in the combustion chambers is fed directly to the crankshaft. Now unfortunately the crankshaft has fitted to one end of it a pulley, and this pulley behaves like a loudspeaker. It was found on this 'structure research engine' that removal of the pulley reduced the noise at 1,600 Hz, its natural frequency, by 25 dB. They therefore constructed an isolated pulley by making two sections, a central bush which fitted on the crankshaft, and an outer, larger-bore pulley, the two being connected by a rubber sleeve. This produced an improvement of 15 dB at 1,600 Hz compared with the conventional pulley.

The result, with extraneous noise sources such as intake and exhaust dealt with and with the gearbox heavily lagged, was an improvement of around 10 dB from about 400 Hz upwards in the research engine compared with its conventional relation. Except at low frequency, where the whole engine vibrates bodily and therefore no amount of damping will help, the effect was to reduce the noise to that of a comparable petrol engine, with the elimination of the objectionable diesel knock.

The other type of engine which was constructed was totally different, and in many ways rather less practicable. A crankcase and cylinder block was cast from magnesium to have walls about 30mm thick. Because of the low density of magnesium, this gave no increase in weight over a conventional engine, but much greater bending stiffness because of the sheer thickness of the walls. There were considerable practical problems because of the low tensile strength of magnesium, and the liability to galvanic corrosion. These were overcome, and the engine was fitted with a highly damped timing cover and rocker cover, isolated sump and crankshaft pulley similar to the structure research engine.

The principle of this engine was twofold. The high bending stiffness of the walls would cut down the magnitude of vibration at low frequency and would give it resonances at high frequency when the Fourier components of the cylinder pressure pulses are small. The resulting engine was better than the structure research engine at low frequency, and very much the same at middle and high frequency.

The principles described in these research engines can of course be applied to an immense variety of other machines. In fact the prevention of resonance is one of the most fruitful design tools when one is trying to build a quiet machine. One way of doing this, as we have seen, is by damping. It is obviously best to gain the required damping by using materials in the first place which have what is known as a low dynamic magnification factor as an inherent characteristic. Rubber is a well-known member of this category, and the qualifications are that the molecular structure of the material must involve long chains of molecules which have to slide over one another under strain, causing frictional energy losses.

Unfortunately, you cannot make many machines out of rubber! However, one of the most significant discoveries of recent years has been that of carbon fibre. Long strands of synthetic polymer are converted by a special heat process so that they are made up of long-chain carbon molecules of high tensile strength. When bonded with resin, these fibres can go to make up a 'miracle' material as strong as steel and a fraction of the weight. The property which has not received much publicity is that the internal damping is very high indeed.

Carbon fibre is still in its infancy and thus expensive. For the moment, economics preclude its use in many applications. Unfortunately, the other materials with high internal damping are seriously lacking in strength or stiffness. This drives one to using conventional metals such as steel, and gaining the damping by external means. We have already seen this done in the structure research diesel engine, but before we go on to apply it to other machines let us just make sure we know how it works.

Figure 52 shows a steel plate to which has been bonded a thickness of damping material, such as synthetic rubber. This plate could radiate noise by vibrating bodily, but more often than not it will have to bend to do so. When it bends, the rubber layer has to bend also, and for reasons of geometry the outside surface of the rubber is extended. If the plate bends the other way

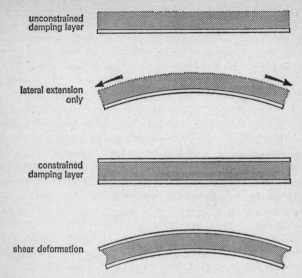

unconstrained
damping layer

lateral extension
only

constrained
damping layer

shear deformation

52. Unconstrained and constrained damping layers

the reverse happens and the outside surface of the rubber has to
contract. Energy dissipation, or damping, is achieved through
frictional losses in the rubber as it extends and contracts longi-
tudinally.

Now let us stick a thin metal sheet to the outside of the rubber
layer. When the plate bends, the outer surface of the rubber cannot
extend or contract because it is stuck to the thin metal sheet,
which has high lateral stiffness. This means that the rubber layer
is severely distorted, with the result that there is a substantial
shear deformation. Anyone knows that to bend a sheet of rubber
is child's play, but to slide one surface any distance against the
other takes a lot of doing. For this reason, damping methods
involving a constraining layer are very much more effective, and
to a certain extent a thin rubber layer is better than a thick one
because there will be greater shear deformation.

Simply gluing a piece or pieces of metal sheet to a resonant
panel will, depending on the size of the sheet, the position, and

the frequencies involved, be very effective. This is not to say that unconstrained damping treatments are no good, and there are many proprietary products on the market which are economical and very easy to apply which provide quite enough damping for many applications. Rubber is by no means the only material one may use. Many substances, from resin mixtures to felt, have their uses. A well-known product looks like porridge and is supplied either in bulk liquid form or as self-adhesive sheet. The latter form can conveniently be bonded to sound-absorbent neoprene foam, thus killing two birds with one cheap and easily applied stone. Alternatively, a couple of layers of glass-fibre matting bonded to steel with resin is almost as good, and slightly lighter and cheaper. It also gives phenomenal stiffness to the steel, much greater than the equivalent weight of either steel or fibreglass. Depending on the resin used, this material can withstand high temperatures.

Nevertheless, the steel/rubber/steel laminate has many advantages as a constrained system. Sometimes it comes as a basic sheet of, say, 20 s.w.g. steel, resin bonded to a lighter skin of about 26 s.w.g. steel, giving an ultimate thickness of around 16 s.w.g. Other products are a multiple build-up of several layers of rubber and aluminium. Unfortunately the bending stiffness of the sheets is poor and fabrication is difficult. It has not yet been possible to make a satisfactory pressing, because one of the layers tends to crinkle or split. Experiments are in progress to try to press the two layers separately and bond them afterwards. When welding the sheet, the 'meat in the sandwich' is locally destroyed and the strength of the weld is poor, because it often forms a bond to only one of the layers of the sandwich.

Damping has innumerable applications in the world of machines generally. The noise from any machine having light metal panels can be cut down, sometimes by 10 dB or more, by applying damping sheet to the panels. More skilfully, if the panels can be designed so that their resonant frequency is nowhere near that of the initial disturbance, this alone can be very effective. A well-known case of damping of panels is that of the quietening of saloon cars and road vehicles in general. Roof panels, floor panels, bonnet and boot lids can have a multitude of resonances and damping brings big rewards. The use of damped bits or 'steels' in road drills can give some 7 dB reduction in noise, providing the exhaust noise has been silenced first. Acoustically the best steel is a hollow one filled with lead shot, but this is highly impractical as heat causes the shot to compress into cubes. Many other types of insert such as copper, manganese or cast iron have been tried, but none quite as good.

Resonance in solids, though, is not the only type of resonance. Remember the Helmholtz cavity resonators from chapter 8. Any enclosed body of air is resonant, and depending on its shape and size it will have one, or the usual chorus, of frequencies. The remedy is simple in theory: either damp the vibrations, for instance with fibre or foam, or design the thing so that the resonant frequencies are out of harm's way. The larger the air volume, the lower the frequency of the fundamental and vice versa. Large cavities can behave like small rooms, and the normal modes described in chapter 10 can be set up. Where there is a connecting 'neck' between the cavity and the world outside, as in a Helmholtz resonator, the larger the neck, the *higher* is the resonant frequency. Do not overlook the fact that a damped cavity resonator is a highly efficient sound absorber, and can be very useful in some cases.

There are, therefore, three lines of attack on a noisy machine. If you cannot smooth out the initial disturbance you can certainly avoid amplifying it through resonance by either damping the resonance or designing it out of the picture. Failing this you can often thwart the radiation mechanism by disconnecting the source of excitation from any other part of the machine as much as possible. Most of it is just applied common sense.

13   Silencing and soundproofing

Sometimes, alas, no amount of information such as that given in the last chapter would be any good to you. If you are confronted with a noise problem caused by an existing machine it is quite useless to go around telling people how the machine should have been designed in the first place. The fact is simple: you are stuck with the machine as it is. Are you stuck with the noise as well?

Imagine a typical case, one which embodies most of the practical problems, a diesel alternator. These can be found in all sorts of unlikely places as standby generators where the whims of the National Grid cannot be tolerated. One of the places where they are frequently found, and where their noise is least welcome, is in hospitals. Let us therefore put ourselves in the position of having to install one of these, without being able to redesign it.

Before anything else is done, an anti-vibrating mounting system must be designed. This will mean fitting the engine and alternator to a common sub-frame, so that they cannot move in relation to one another as a result of the couple force between them. Resilient mountings, either springs with rubber pads or rubber-in-shear units, will then be used, perhaps six or eight in number. These must have the necessary natural frequencies in both vertical and horizontal modes to cope with the lowest forcing frequency likely, remembering that several of the six modes of vibration of the generator may be coupled.

However, it is not enough just to sit the generator on its mountings and hope for the best. All connexions to either engine or alternator must be flexible. This includes fuel pipes, conduits and air ducts. If anti-vibration mountings were not used, subsequent silencing measures would be severely affected because the ground or the building for long distances from the generator would be just as much a noise source as the generator (remember the tuning fork), so that however marvellous a soundproof room one built round the machine a large and important part of the noise source would still be unenclosed.

The heart of the problem is, of course, the enclosure. The simplest way of stopping noise going from A to B is to block its path. If a source radiates in all directions it will have to be boxed in. Chapter 9 told us about sound insulation, and the first thing we need is mass. If we can build a brick or concrete enclosure, all well and good, and, as Appendix 2 shows, the transmission loss of the walls will be very good. This might not be very practical though. We might not be able to stand the weight or might want to be able to dismantle the enclosure, in which case we have two alternatives. We can sacrifice some noise reduction and go for a lighter material for a single-skin wall. Purely on mass, 16 s.w.g. mild steel sheet or 25mm of dense chipboard is economical and readily available. However, steel has such low internal damping, because it is stiff with small inherent damping, that it would be very resonant. We know that resonance punches holes in the mass law. Some external means of damping will have to be added. Before we choose this, let us consider the next problem.

Remember that in chapter 10 we found that if you build a room round a noise source you do not just subtract the transmission loss of the walls from the number of decibels you started with and arrive at the final sound level. You have to add ten times the logarithm of the area of enclosure facing the direction of interest, and subtract ten times the logarithm of the average absorption coefficient of the surfaces inside the room. This is because if a single sound wave travels from the source to the wall of the enclosure, part of it will pass through and part will be reflected back. The part reflected back will be reflected again and strike the wall again, and another part of the original wave will get out and so on. In other words the part of the original sound wave which does not win its battle to penetrate the wall is reflected and lives to fight another day. Therefore, if the walls are reflective inside, more sound will escape than if they are absorbent.

This means that the enclosure walls must be acoustically lined. If they are not, and the average absorption coefficient is as low as, say, 0.05, then $10 \log_{10} 0.05 = -13$ dB must be added.

The effect of the enclosure is reduced by 13 dB. Let us therefore line the enclosure with 50mm of rock wool, and allowing for the reflective surface of the diesel alternator itself 'guesstimate' the average absorption coefficient at 500 Hz for instance as 0.6. $10 \log_{10} 0.6 = -2$ dB, so we are 11 dB better off. (All the calculations should be done at least for each of six if not eight octave bands.)

There will be a practical problem in that the rock wool will not stay up on its own, but this can be overcome by using anything from wire netting to expanded metal, or, as with acoustic tiles, using perforated metal. With the sheet metal enclosure there is a big bonus because the mineral wool retained against the metal sheet will provide the required damping. With timber, damping is not quite so necessary because there is a fair amount inherent in the material.

If one cannot go to heavy masonry but needs more attenuation than a single-skin 16 s.w.g. steel panel will give, it is of course perfectly possible to apply the principles of using two isolated skins, as in chapter 9. One way of doing this is to start again and build another enclosure of the same construction outside the original steel sheet and mineral wool one, leaving a gap of 150mm or so between and mounting both enclosures on resilient sponge rubber strip. (It is well worth doing this even if you are stopping at one enclosure.)

A more sophisticated system involves using lighter membranes sandwiched between and supported by layers of mineral wool. There are, of course, many alternative sound absorbents which may be used, neoprene foam and polyurethane foam being examples. These, however, have to be specially treated to avoid a fire hazard, though they are usually more economical and do not require any retaining mechanism. Any old plastic foam will not do. It must have inter-connecting cells, and have a high enough flow resistance.

The physical dimensions of the enclosure will affect its performance in two ways. Firstly, if one is interested in radiation in a particular direction, a large wall will radiate more sound than a small one. In addition, as we are interested in the average absorption coefficient inside and as some machines themselves present a large area of hard, reflecting surface, for the same machine and the same lining the average absorption coefficient inside a large enclosure will be greater than for a small one. It is thus often best to use a fairly large enclosure, particularly because you can then put a door in it and walk in for inspection and maintenance purposes, instead of having to fit an array of access panels.

Typical examples where air resonance reduction can reduce noise include the flue of a boiler, where pure tone components can be added to an otherwise random noise because the random initial disturbance caused by the combustion of fuel and air behaves like the player blowing into the recorder mouthpiece in chapter 3, and the exhaust stack then behaves like a resonant tube. It can be difficult to avoid this – it is a question of careful sizing of dimensions. Every boiler installation is different, so one cannot give a general recommendation, but if stack resonance is troublesome a silencer of the kind described in the next chapter can sometimes be the only way out.

Another important example is that of a fan with an intake or discharge duct. It is obviously most important for the duct not to be resonant (like a tube) at the same frequency as the blade-pass note or harmonics of the fan. This can happen in a vacuum cleaner or an electric motor.

Reducing resonance can prevent unnecessary, and sometimes substantial, amplification of a noise, but often more fruitful is the reduction of radiation efficiency. Remember that every point on a radiating surface can be regarded as a separate point source of sound. This has two effects: firstly, it causes destructive interference at the edges of surfaces (see figure 31), but secondly it means that a large surface which is twice the size of a smaller one and vibrates with the same amplitude will radiate twice the sound power, or 3 dB more. When you strike a tuning fork and hold it in your hand, it is barely audible because of interference between the prongs and because it has such a tiny surface area. If you then press it against a table its sound level goes up surprisingly. This is because the vibrating energy is fed to the table which is forced into vibration by the fork, and because it is a large flat surface there is less interference and a large area. In engineering calculations for small dimensions, one can usually ignore the fact that a certain

amount of power is absorbed by the table and say that a tuning
fork on a 0·01 m² board will be 3 dB quieter than one of 0·02 m².

Therefore if you have an electric motor, say, bolted rigidly to
a floor, you would have to count the floor as part of the source,
and even if resonance does not have a look-in, the radiation of
motor noise will be highly efficient because of the sheer size of the
surface involved. The simple remedy of placing the motor on
resilient mountings will secure a sizeable reduction in noise. Bearing in mind that the floor is connected to the rest of the building
and that in some materials, such as concrete, sound will travel
long distances without much loss of energy, you will realize that
there is the additional hazard of extending the effective 'noise
source' into quite remote parts of a building unless the machine
is properly isolated.

Even on a smaller scale, this question of radiation efficiency is
of first importance. I recently saw a pipe which vibrated strongly
at 3 KHz. It had been well lagged and radiated nothing significant
itself, but the angle-iron supporting brackets clamped to it, which
were outside the lagging, were themselves screaming and causing
the whole structure to emit this note.

The effect of putting a typewriter on a felt pad is well known,
but strangely enough few people have the sense to do the same to
other machines. Office workers often have to endure quite unnecessary thumping from addressing machines simply because
they are sitting solidly on a large surface. Much more serious is
the effect when a 200-ton press is bolted directly to a floor slab,
when occupants of the next-door building will suffer. Pumps can
transmit noise into piping which in turn is fixed to walls, and even
motors on flexible mountings can transmit noise down solid
electrical conduit. When I bought my car the electric screen
washer made an abominable noise simply because it was fixed
direct to the bulkhead. Resilient mounting immediately made it
inaudible.

It is almost true to say that before you do anything else the seat
of the initial disturbance should whenever possible be isolated
from everything else. This is particularly important on large
machines to which are fixed pumps and motors, and sources of
impacts. There are usually resonant conditions as well, and of

course isolation of the source will greatly reduce the need for damping.

What, then, is exactly meant by vibration isolation? It is no good just bringing any old piece of cork or rubber under a machine; it could either make no difference at all, or even amplify the vibrations to an alarming extent. Most isolators are designed to deal with low-frequency vibration, even below the audio range. It is unfortunately beyond the scope of this book to go into the subject of vibration isolation in depth.

Once again, we are trying to achieve an impedance mis-match, and this means the mountings must be as resilient as possible; the lower the stiffness, the better the isolation. However, the mountings of a machine have got to be massive enough to take the weight of the machine. The mountings may take the form of springs, either helical or leaf, or other spring-like materials such as rubber, or even air. Now unfortunately a spring with a mass fixed to it will be resonant at a certain frequency, and if the exciting force in the mass has the same frequency you will of course have amplification, which is only controllable by means of damping. Even if the damping is so great that when the spring is depressed it just creeps back to its equilibrium position, you will have no isolation of vibration and you will just be avoiding making things worse.

In fact, you do not get any improvement until the forcing frequency is over $\sqrt{2}$ times the natural frequency of the mounting under load, and nothing useful until the ratio is over 2. In theory, the isolation efficiency goes on improving as the ratio increases, but this assumes many things, including a perfectly rigid base. In practice one cannot count on more than about 20 dB isolation, and if the base or floor is very mobile the benefit can be much less, particularly at resonant frequencies of the floor. Springs on their own cannot be used for isolation of audio frequencies. This is because high-frequency sound will happily travel in the metal down the coils of the spring. In many cases where low-frequency vibration is the only concern this does not matter, but whenever audio frequencies are involved rubber pads should be incorporated in the mountings, or where possible rubber mountings should be used.

Once again it is essential to realize that it is useless to use dense rubber in compression except when the lateral dimensions of the rubber block or strip are small compared with the thickness. Nearly all rubber mountings (a typical one is shown in figure 53)

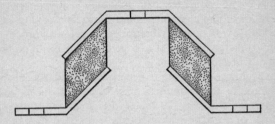

53. A typical rubber anti-vibration mounting (section)

use rubber in shear, so that it is much more resilient.

The natural frequency of a mounting is related to the amount it deflects under the applied load. Figure 54 shows the relation.

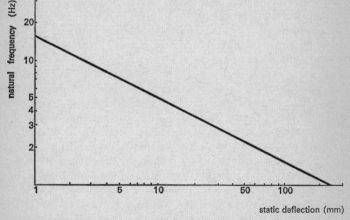

54. Graph showing the relationship between the frequency of a mounting and the amount of deflection under a load

However, the calculation of the forcing frequencies and the required type of mounting must be left to an expert in all but the simplest cases. A machine can vibrate in six different ways at six different frequencies all at once.

Great care is needed in the construction of doors and panels. A 2mm gap around a personnel door in a 4m^2 wall can knock 15 dB off the transmission loss (figure 43). The doors must have a transmission loss themselves of within 5–10 dB of the main structure and have airtight seals. This means using foam rubber or phosphor-bronze sealing strip and a pressure fastening. As far as windows are concerned, if the area is not too great, a single pane of 6mm plate glass is best, but a second pane may be added if both are resiliently mounted.

However, if you remember that what we are enclosing is a diesel alternator, there are going to be some problems, of which cooling can be quite a serious one. It may be necessary to pass several thousand cubic metres of air per hour in and out of the enclosure if the engine radiator is part of the set. How do you let air in and out without letting sound out?

The first necessity is to cut an inlet and an outlet aperture in the wall. This will cope with the air and let out a blast of noise. As usual, the hole can be regarded as a sound source in its own right and it will radiate sound hemispherically. Interference effects will come into play with increasing frequency, causing the holes to radiate most high-frequency noise directly in front of them. What can be done about it? What can ever be done to silence sound travelling in a gas stream along a pipe or duct?

If a duct is fitted to the aperture, sound will then travel along the duct in two ways. Some sound waves will enter the duct obliquely, and will travel along it by being reflected from side to side. Other waves will travel straight down the duct as plane waves without touching the sides. Obviously if the walls of the duct are not reflective – that is, they are absorbent – the first type of wave will not get very far. How far they will get will depend on their angle, the duct width and the absorption coefficient of the duct lining. With an average type of sound-absorbent lining the waves incident at angles greater than 30° will be reduced to the level of the plane-axial waves within a length of about four duct widths (figure 55).

As far as the plane-axial waves are concerned, the reasons for

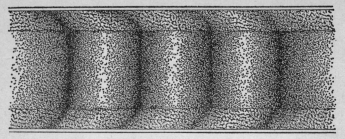

55. Plane-axial waves in a lined duct

their attenuation are not so simple. As a plane wave travels along a lined duct, the edges of the wave will travel in the sound-absorbent lining. Viscous drag in the pores of the fibre will tend to dissipate that part of the sound wave as the air particles oscillate to and fro. However, in accordance with Huygens' wave construction, the energy which is dissipated at the edges of the waves is soon replaced from the main body of the wave, in a similar way to that in which the sound shadow beyond a screen is encroached upon by the diffraction process. The net result is that the dissipation at the edges saps the energy of the main part of the wave and attenuates it as it travels along the duct. In addition, the velocity of sound in the sound-absorbent lining is much lower than in ordinary air – around 200m/sec compared with 334m/sec. This has the same effect as a heat gradient, and 'bends' the wave sharply back at the edges as the part of the wave in the lining trails behind. This bending of the wave front changes the direction of travel of the wave at the edges down into the lining. As the lining is not totally absorbent the wave will be partially reflected back across the duct, but very soon be virtually completely absorbed. Meanwhile this bending at the edges, which is equivalent to an expansion of the area of the wave front, means that the intensity of the plane wave is further reduced. In just the same way it is the increase in area over which a spherical wave has to stretch which accounts for the inverse-square law.

Four factors govern the degree by which the plane wave is attenuated: the width of the duct, the thickness of the lining, the flow resistance of the lining and the frequency of the sound. The

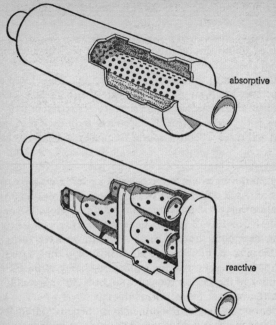

absorptive

reactive

57. Exhaust silencers

The expansion chamber will be connected via a pipe to the source, and via a tailpipe to the atmosphere. The source, if it is an engine exhaust, for instance, will send periodic pulses of air, made up of a simple series of sine waves with a large concentration of energy at the fundamental frequency, down towards the expansion chamber. This is analogous to the exciting force (p. 211) in the vibrating machine on a spring. The expansion chamber is an enclosed body of air, and has a particular resonant frequency like any cavity resonator. If the resonant frequency coincides with the fundamental frequency of the source, you will have the same effect as if the vibrating machine is put on a spring of the same natural frequency: large amplification. The tailpipe is a resonant tube, and this has a resonant frequency which will make the analogy equivalent to putting the machine and spring on a springy floor. The effect will be to give a combined funda-mental resonant frequency to the cavity and tailpipe. If the two frequencies coincide, amplification will be very large, and trans-mission of sound through the silencer will only be governed by frictional losses during the gas oscillations.

Turning again to the machine on the spring, we see that as the machine speeds up, amplification decreases, and there comes a point where mechanical isolation starts to occur. Exactly the same happens with the resonator. If the source frequency rises, ampli-fication will cease when the source frequency is $\sqrt{2}$ times as high as the cavity fundamental resonant frequency. As the ratio increases, air vibrations are less and less successful in getting through, until we reach the frequency of the second harmonic resonance. As the frequency of the source rises still further, the attenuation curve will increase again but dip violently down again at higher harmonics of the resonant frequency.

When designing the resonator silencer, we have three main things to think of. First there is the resilience of the 'spring'. The larger the cross-section of the expansion chamber the softer the air spring and the greater the potential isolation. Next in import-ance is the natural frequency of the chamber, which is largely dependent on its length. Finally there is the natural frequency of the tailpipe, which is dependent on its diameter and length. There are, however, two more factors which can throw one's predictions out, and these can be very difficult indeed to calculate, namely the source impedance and the terminal impedance.

Let us take the source impedance first. With the vibrating machine on a spring, no matter what mounting you place the machine on, you will not alter the force in the vibration very much. In other words the source impedance is high, in just the same way as the silencing system will not greatly affect the exhaust pulses from the engine provided the back pressure is small. The terminal impedance is also high because of the suddenness with which the sound wave meets the large unrestrained air mass outside (re-member that this is what makes a tube resonant because of the reflection which occurs). On the other hand, the cavity impedance is low, and it is the mis-match caused which gives the attenuation. Therefore, if the source or terminal impedances are altered, the very same silencer will not have the same efficiency. A typical source with a low impedance is a loudspeaker, and consequently if you test a reactive silencer using a loudspeaker for a source you may get unduly pessimistic results. Equally so, if you do something to the tailpipe, such as connect it to another silencer, you may lower the performance of the original system by lowering the terminal impedance. In the mechanical system, this is the reason why you will get better isolation if the spring sits on a massive base than if it sits on a soft or light base.

Of course reactive silencers do not stop at single straight-through cavities. The cavity may be a side branch like a resonator absorber (chapter 8). Now the reverse happens to the case of the straight-through expansion chamber. The greatest, rather than the least, attenuation occurs at resonance, where energy dissipation takes place during the highly amplified oscillation of gas in the neck of the resonator.

In practice, exhaust silencers are multiple-expansion chambers with the occasional side-branch resonator thrown in. One important factor is that if the connecting pipes protrude into the chamber attenuation is improved by making the 'air spring' comparatively softer. The placing of two chambers in series does not give double the attenuation necessarily, but with the right connexion between them the effect can be better than double. Most reactive silencers on the market have been evolved empirically many years ago and still the method of research seems to be 'suck it and see'. It is to be hoped that great use is now being made of analogue computers in the design of really efficient silencers.

The snag to these silencers is that they tend to have frequencies at which their performance is very poor, and they are much affected by the surrounding installation. The harmonics of resonances come thick and fast at high frequency and their effect can be poor. Many silencers thus incorporate an absorption stage as well to cope with high frequencies, and a very good system for an engine is to install first a reactive silencer and then farther down the pipe to fit a straight-through absorption silencer. One silencer manufacturer recommends that best results are obtained with a tailpipe length equal to ten times its diameter.

Combustion engine exhausts are not the only application for

silencers, either reactive or dissipative. A very fruitful use for both types is on the exhaust port of a pneumatic drill, but, unfortunately, however well designed the silencer there is always a certain penalty in reduced efficiency. The order of reduction obtained is 5–7 dB, which is not startling because exhaust noise is only one of the sources. Parallel use of a damped steel would improve matters further, and damping and insulation of the casing help. A common type of silencer fitted to road drills is the muffler jacket, which serves two purposes. It consists of an impervious loose sleeve or bag which fits over the exhaust ports and is laced to the drill at the bottom, where there is a ported disc to allow the exhaust air finally to escape. The whole bag is lined with absorbent material and thus forms a partially lined duct with the casing of the drill. In addition, by shrouding the drill, the bag has a small effect on the radiation of casing noise. However, whatever you do to a drill, you are always limited by the fact that the steel is in solid contact with the road, and by means of the tuning-fork effect the road becomes a radiator of noise.

We can now go back to the diesel alternator and fit one of these systems to the exhaust (taking care to use a flexible connexion). We can also use an air-intake silencer and filter combination working on the same principles. It is important, though, that reactive silencers radiate noise from their casing, and as we may have to site ours outside the acoustic enclosure this could spoil everything. The simplest solution is to bury the silencer under as much earth as possible. Otherwise the silencer may have to be lagged.

There are two ways of lagging things, and it is very important to use the right method. If the casing of the item to be lagged is in resonance, and this can be discovered by testing its response to vibration of different frequencies and comparing the resulting spectrum with the spectrum of the noise under normal operation (a blunt instrument and a good ear is sometimes adequate, but not to be relied on), then damping lagging may be all that is needed. A suitably thick coat of damping compound, such as one of the resin-based proprietary products, will greatly reduce resonance. However, with many things such as silencers, the casing vibration is forced upon it by what is happening inside and damping will not affect it by more than the odd decibel. If resonance is not the case, simply doubling the mass of the casing would improve matters by about 5 dB, but this is not very rewarding.

The best thing to do is to invoke the principles of chapter 9 and add a resiliently supported impervious skin of about 5 Kg/m^2. The resilient support can be achieved with a layer of mineral wool, which will also damp what resonances there may be. The outer skin, which must not make solid contact anywhere with the vibrating surface, can be anything from metal sheet to hard-setting compound or one of the proprietary high-density flexible plastic sheets.

To digress a little, this problem often occurs when valves in a high-velocity gas pipe system cause violent turbulence downstream, because gas velocity through the actual valve orifice exceeds Mach 1, the speed of sound. Vortex formation occurs on an immense scale, and if there is an appreciable length of pipe it will radiate searing high-frequency random noise. It has been found that with this type of very high-frequency noise, thick lagging of dense porous fibre, such as high-density fibreglass, is effective without an outer skin. This goes against the folklore of acoustics, but has been found to be true in several recent cases. The Institute of Applied Physics in Delft are carrying out a research programme on pipe noise, and we can expect some interesting results. However, pipe lagging in these cases is not always the only necessary solution, because often pipe vibration by-passes the lagging via supporting brackets and radiates from the main structure. It is the old story of the tuning fork and the table, and resilient support brackets have to be used.

Let us now just put the finishing touches to the diesel alternator scheme. First of all, there must be no gaps or cracks in the enclosure walls. Secondly, the enclosure must on no account make solid contact with anything vibrating, and this includes the exhaust system, which must be supported from the ground and only flexibly connected to the engine; and the engine/attenuator unit itself must not make solid contact with the ground. Where the exhaust pipe passes through the enclosure wall, an oversize hole must be made and a flexible grommet of asbestos fabric used to seal the hole. For other pipes which are not hot, rubber or felt may be used. A little extra improvement will be gained if the exhaust tailpipe and cooling air-silencer louvres do not point in a critical direction because they will be directional at middle and high frequency.

Unfortunately there are many machines where one simply cannot tolerate total enclosure for practical reasons. For one thing, such frequent access may be required to part of the machine that even a power-operated access door is impractical. Examples are small presses and riveters which the operators are constantly feeding with workpieces. An enclosure with a hole in it will not be useless. The intensity of the sound emitting from a hole is logarithmically proportional to its area, and, for reasons of impedances again, when the wavelength of sound is large compared with the hole, reflection occurs, just as when a plane wave arrives at the end of a tube. Thus low-frequency sound is not very good at getting out of relatively small holes. In addition, when it does get out, it is almost non-directional and you do not get the intense beaming of sound to the front of the hole that occurs with sound whose wavelength is short compared with the size of the hole.

If the aperture is necessary, for instance to admit a conveyor belt to the machine, the problem can often be solved by building an absorbent-lined tunnel along the conveyor which will act like a lined duct or straight-through silencer. The whole enclosure can sometimes take the form of a lined tunnel, although if its dimensions are large it will emit a lot of noise from the ends.

Do not forget that these machines are all usually in rooms, and we have already gone into the question of reverberation in rooms. Many factories do not contain much in the way of sound absorption, and so a couple of metres from a source reverberant sound engulfs you. This is most important where directional enclosures and screens are concerned, because you get an improvement only if the observation point is close enough to the machine for direct noise to have predominated before the erection of the screen. If there is only one major noise source in the room, screening is more effective, because if the screen is very close to the source it will cut down the sound power output of the machine/screen combination by absorbing some sound before it has been able to travel any distance. This is particularly true if the source is very directional and the screen is situated in that direction. However, remember the ubiquitous logarithms. If a source is non-directional, and a screen encompasses it around 180°, even if the absorption coefficient of the screen is $1 \cdot 0$ – that is, total absorption – the sound power fed to the room as a whole will be knocked down by only 3 dB – barely noticeable. To make it worse, the screen will probably not have an absorption coefficient as high as $1 \cdot 0$ and it will only remove energy, for a non-directional source, equivalent to ten times the logarithm of the proportion of machine it encompasses multiplied by the absorption coefficient of the screen:

$$\text{Reduction in sound power} = 10 \log_{10} \frac{(1 - \theta\alpha)}{2\pi}$$

where θ is the angle in radians of the segment shielded by the screen and α is the normal incidence absorption coefficient of the screen. Therefore if the screen extends halfway round the machine and has an absorption coefficient of only $0 \cdot 7$, it will reduce reverberation by only 2 dB. You can extend the maths further if you know the directivity of a directional source and things can be a little better, but never startling.

This screening will help direct noise but remember that sound diffracts over a screen, with increasing effect at lower frequencies (figure 34). Making the screen highly absorbent on the source side will reduce diffraction a little, but it is unwise in rooms to count on more than 5 to 6 dB screening above 500 Hz, with nothing much below. The effect varies considerably with positioning of the source and receiver because this affects the angle into the sound shadow and the effective screen height. Also the sound which emerges at the edge of the screen is virtually a long line of sources, and sound from the two sides will give rise to interference patterns (as in figure 35) even for random noise, and cause lobes of good screening and poor screening.

A sure-fire way of reducing reverberant sound is of course to raise the average absorption coefficient of surfaces in the room, the effect being indicated by figure 44. It is of paramount importance, though, to grasp that you have in most cases to *double* the total absorption before you will even notice the difference. Putting acoustic tiles on the underside of a factory roof usually only makes 5–6 dB difference, and the machine operators are often in the field of direct sound from their machines, which of course is totally unaffected. There are a few tricks to improve things, one of which is to fill in roof trusses with sound-absorbent panels. There are available functional absorbers, large absorbent bodies which if hung from the ceiling in large enough quantities can sometimes overcome difficulties of negotiating labyrinths of pipework and ducting with acoustic tiles. It is good policy to concentrate as much absorbent treatment as possible in the neighbourhood of sources.

However, reduction of reverberation by increased absorption is a most expensive means of achieving only a small reduction. It is always the best policy to start by buying the quietest machines, and even to take expert advice and modify machines wherever possible so that they emit less noise in the first place. Anti-vibration mounting throughout a factory will stop the tuning-fork effect and sometimes do more than an acoustic ceiling, particularly at low frequency. Remember that springs alone are poor isolators at high audio-frequencies. If machine modification is impossible, well-designed acoustic enclosures can in most cases bring about plenty of noise reduction, even when some apertures are essential. It often requires no little skill to design an enclosure-cum-screen which reconciles acoustics with ergonomics and economics, and fortunately there are several very experienced specialist manufacturers of this equipment.

There are, of course, cases where people do not much care about the noise in the area where it is being made, but other people are vociferous on the subject. Many a factory annoys its neighbouring residents, and many a hi-fi enthusiast has driven his neighbour to distraction. As far as factories are concerned, obviously all the recommendations on the subject of internal quietening would result in a lowering of noise outside as well. However, sometimes there is absolutely no need to worry about the internal noise level, and improvements to the building are called for. This certainly applies to the hi-fi man.

He will not work wonders by slapping acoustic tiles on the

party wall. We saw in chapter 9 that this will not increase the transmission loss of the wall. The chances are that his room is fairly absorbent anyway with its carpet, curtains and well-upholstered chairs. Acoustic tiles on the party wall would do no more than double the absorption and this only improves matters by 3 dB. If he clads all four walls and the ceiling with tiles he might squeeze a few more decibels, but penetration of direct noise through the party wall would take over from the now diminished reverberant sound, and no amount of absorption affects direct sound.

Unless the room involved is very reverberant, with an average absorption coefficient of below 0·1, sound absorption is not the answer to a problem of room-to-room transmission of noise, unless – and it would be rare – a handful of decibels will suffice. The problem is by no means an easy one to solve, and even if the hi-fi man doubled the mass of the party wall he would at most get 5 dB improvement. He has got to turn to the resiliently supported second skin. He will have to fix soft rubber isolators, as few as possible, to his wall, and to them fix an impervious membrane of some $10Kg/m^2$ which nowhere makes solid contact with the structural wall. The intervening space should be as wide as possible, certainly not less than 50mm, and be filled with a sound-absorbent blanket of, for instance, mineral wool, to stop reson-ance of the enclosed air mass. The edges of the membrane should be sealed to the structural side walls and ceiling with mastic; again there must be no solid contact.

To be a kill-joy once again, I must say that even this method will not have its full effect because sound will enter the structure as it impinges on the other walls, travel into the adjoining room in the structure and be re-radiated as airborne sound from the walls of the near-by room. In a block of flats, sound may travel some distance in the concrete and cause widespread trouble.

When structure-borne noise is a really serious problem, just about the only solution is the construction of a room-within-a-room of the sort described in chapter 9, itself mounted on isolators, and nowhere making contact with the structure. The space be-tween the inner and outer rooms is absorbent-lined, and each room has a soundproof door.

Nevertheless, there are many cases where the special construction of false walls and ceiling can improve transmission losses by over 10 decibels, even allowing for flanking transmission. A variation on this theme occurs in the floating floor, which both improves the transmission loss of the floor/ceiling structure and also reduces impact noise. This latter type of noise is sometimes just as important and is commonly experienced with footfalls annoying people in the room below. Soft floor covering helps, but floating floors have many advantages, one of them being thermal insulation particularly with underfloor heating.

Floating-floor construction consists of mineral wool quilt laid over joists or a concrete floor slab, with either a boarded floor or a thin concrete screed placed over it. Several points are important: with wooden joist floors the quilt must not be overloaded so that it is compressed solid, and nails must not penetrate from the floating floor to the joists. With concrete screeds, there is a limit to the area of floor which can be tolerated without causing a liability to crack, and for large areas subdivisions are necessary. In all cases the quilt must be turned up at the edges and the skirting board if present must stop short of the floor. As usual, the maxim is 'no solid connexions'.

This sort of thing does not often apply to factories, where the roof is often of pitched corrugated-asbestos construction. If noise penetration of the roof is a serious thing, an *impervious* suspended ceiling must be installed at as low a height as possible. A plain acoustic suspended ceiling is not suitable unless given an impervious backing. Some fibre acoustic tiles are naturally impervious and are ideal in that you get an added bonus of a lowered reverberation level. Otherwise, a suspended plasterboard ceiling will improve the transmission loss, without of course affecting the reverberation. The suspension rods should be kept to the minimum, and ideally should be fixed resiliently.

Fortunately, factory roofs are not usually one of the weakest points; these are often doors and windows. It sounds silly to say 'keep them shut' but it is often the remedy. However, though the public health inspector will now stop complaining of external noise, the factory inspector will start shouting for ventilation. A simple air-extraction system can be installed, but of course the air outlets and inlets must have absorption silencers. If the factory noise is a real humdinger and shutting windows and doors is not enough, the windows can be filled with glass bricks, and special doors of heavy construction with airtight seals may be necessary. This can bring great problems if there is a lot of traffic through the doors, although partial relief can be had by acoustic screening of the door aperture instead. On top of this, workpeople may feel claustrophobic with closed-up windows even if the place is fully air-conditioned. In cases like these it is as well to think again about reducing the noise at source.

If all else fails, as a last resort you can always stick your fingers in your ears. This may hamper your working capacity, so fortunately there are ways of achieving the same thing while leaving the hands free to do other things. Ear protection has great advantages, but even greater disadvantages. Four forms of it exist, two of them internal, two external.

The first of the internal varieties consists of disposable wads or plugs of flexible material. Remember that all the principles of acoustics cannot go by the board when it comes to ear protection, and so it is no good putting cotton wool in your ears. This is too porous and wanting in mass. In an emergency, cotton wool soaked in liquid will help, but is obviously not generally practicable. Many chemists sell little balls of waxed fibre which are kneaded into the shape of a plug, but these are unpleasantly waxy. Undoubtedly, the best form of disposable ear protection is a product known as glass down. Glass fibre, so fine that it does not have the irritant properties of normal fibreglass, is supplied like cotton wool, often in a dispenser, and pieces of it are rolled up and inserted in the ear.

The other type of internal ear protector is the moulded plastic ear plug. These are made in a number of sizes, and except with a few 'universal' designs correct fitting is essential. Regular sterilization is also essential. Some ear plugs are downright uncomfortable, but others are much better and can be worn all day. An ingenious design incorporates a 'valve' which causes the plug to close only when the sound level goes over a set limit, so that in a fluctuating noise level there is less temptation to remove the plugs in a quieter period and not put them back.

External ear defenders, or ear muffs, are in some ways preferable to ear plugs. It is easier to see that they are worn, and hygiene is less of a problem. They are sometimes superior in performance, and consist of an outer shell of plastic, a foam lining and a fluid seal; the seals can be interchanged quite easily. The problem with these is often one of heat, and also any sort of particles in the atmosphere can get underneath and cause irritation.

For very intense noises, much over 130 dB, ear defenders are not adequate because vibration of the skull both transmits sound to the inner ear, and is itself very unpleasant. I once measured the noise from a racing engine in a test cell which reached 150 dB, and was wearing only ear defenders. Resonance of bones and cavities in the face was extremely unpleasant, and even thinking was difficult (more so than usual, I mean). In a situation like this, at least a complete helmet is required, and even then exposure should not be prolonged.

It is, however, very difficult indeed in some cases to achieve success with ear protection. Part of the reason is that workpeople would rather suffer the noise than the discomfort or inconvenience of wearing ear defenders, and do not really believe in or care about hearing damage. Unfortunately, though, in some spheres there is a feeling that there is something almost heroic in working in noisy conditions and older members of the community boast of their hearing loss. This leads to a feeling of cissiness in wearing ear protection. Because of the physical and psychological difficulties, therefore, employers must not be content merely to purchase the requisite number of ear muffs, hand them out and sit back complacently feeling they have done their duty. At least a programme of education and preferably the introduction of an organized hearing conservation scheme is required.

In many cases the wearing of ear protection improves communication by lowering the level of the jumbled signal reaching the ear into an area where the analysing mechanism in the brain is better able to single out the meaningful sounds. On the other hand, there is the danger that warning signs, for instance of impending breakdown of a machine, may go unheeded and possibly lead to accidents.

It must be concluded that ear plugs or ear muffs are the answer only when either the wearer's subjection to the noise is for a relatively small proportion of the time the machine is running (particularly if only a few people are subjected to the noise anyway, for instance maintenance staff visiting a largely automatic process) or when other means of noise control are totally impossible for practical and economic reasons.

14 Diagnosis and prescription

No matter what array of medicaments fills your acoustic medicine chest, they will all be useless, and some worse than useless, unless you know when to prescribe them. Perhaps the only fatality likely as a result of wrong acoustic diagnosis is that of the managing director who splutters his last when he hears of the total waste of tens of thousands of pounds. Nevertheless, 'a little learning is a dangerous thing' and if you slap this book shut and plunge headlong into a knotty problem of your own, take care.

This chapter could be filled with sorry tales of men who did not know the difference between sound absorption and insulation, or thermal and acoustic insulation. This confusion is not helped by the makers of thermal insulation who sometimes get carried away and call it 'thermal and acoustic insulation' when what they mean is 'thermal insulation and acoustic absorption'.

Diagnostic techniques in acoustics call for far greater skills than knowing the difference between insulation and absorption. As we live in an age when much acoustic treatment is carried out as an afterthought, the acoustical engineer often arrives at the scene of a noise problem and is told 'there is the problem, what do we do?' Let me take some typical examples from my own casebook.

A factory was in the process of installing five large machines for drying gelatine. Each consisted of a tunnel along which granules of gelatine were conveyed. At intervals along the tunnel hot air was constantly blown upwards through the gelatine, the blast provided at each point by a large centrifugal fan. The air was then exhausted vertically at each of the six points, but because it was laden with gelatine particles it had to pass through a separation system. This consisted of a cyclone, by which means the air was spun via a scroll into a cone-shaped vessel, and centrifugal force caused the granules to impinge on the skin of the cyclone and fall to the bottom, where a motorized valve re-admitted the gelatine to the drier.

At the time of the survey, only two driers, thus twelve cyclones, had been commissioned. Hygiene requirements had resulted in hard reflective surfaces on the walls, and a reflective false ceiling had been installed. The highest noise level recorded, on a platform between the two driers, was 104 dBA, and on other platforms, where men worked regularly, 97–8 dBA. An octave-band analysis revealed a very flat spectrum, as figure 58 shows, well exceeding any hearing-damage risk criterion for an eight-hour continuous exposure. Ear defenders were available to the staff.

The noise was a high-pitched broad-band hiss, and did not appear to come from any particular direction because the reverberant field predominated everywhere. Measurements of vibration

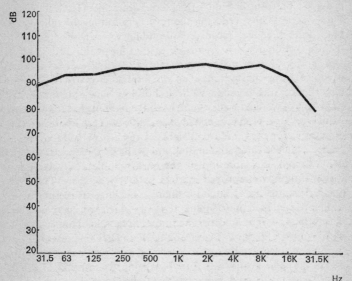

58. An octave-band analysis of the noise produced in the case of the two driers mentioned in the text

were made, using an accelerometer, on a number of surfaces, and it soon became clear that the initial disturbance was the gelatine granules, which harden like gravel, striking the walls of the cyclones. Vibration of the cyclones, of the same frequency make-up as the noise, dwarfed all readings in any other part of the plant. Even allowing for the smaller area of the cyclones to the drier bodies, it was clear that significantly more noise was actually radiated from the cyclones than from the driers to which they were solidly connected. The tuning-fork effect did not mean that the initial disturbance was not the seat of the greatest radiation. The noise from the fans, which had intakes ducted from outside, was not serious.

Because of the strict hygiene requirements (the whole area was hosed down with very hot water regularly) it would have been very difficult to devise a method of sound-absorbent treatment for the surfaces of the building which would stand up to this, and not provide a breeding ground for bacteria, quite apart from the cost. Purely acoustically, reducing the reverberation level would have been quite effective, because the building was so reflective that the absorption could be increased many times without difficulty.

The only solution, in the circumstances, was to prevent radiation of noise from the cyclones, as no modification to the initial disturbance was possible without building new cyclones with some sort of resilient walls. Applying only damping to the cyclones would not have been very effective, and this is a case where correct diagnosis crops up. Although stainless steel, of which they were made, has little internal damping, the walls of the cyclones were being forced into vibration by the hard granules, and damping would only have improved matters by a few decibels. The best answer was to apply sound-insulating lagging, consisting of a resilient layer of foam or mineral wool, which would give some damping as a bonus, and an outer impervious skin of any material with a surface density of over about $3Kg/m^2$. This presented no hygiene problems so long as the foam or wool was well sealed up, and in the end a proprietary form of lagging, foam-backed loaded P.V.C. sheet, was used, being relatively easy to apply.

Another rather different problem affected some people whose office opened into a light well in the centre of the building. Air-conditioning had been installed in a restaurant whose kitchen also looked on to the well, and five condensers had been sited in the bottom of the well. The condensers were cylindrical units each with a small compressor and heat exchanger, but a 600mm diameter propeller fan was the main source of noise on each one.

Figure 59 shows the octave-band analysis in the office concerned,

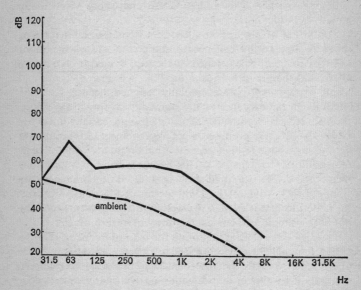

59. An octave-band analysis of the noise produced in the case of the office mentioned in the text

with the windows open. Notice the sharp peak at 63 Hz, which made the noise sound like a piston-engined aircraft, quite uncharacteristic of this type of noise source. The noise level without the condensers running was very much lower.

The light well was some 12 metres long, 2·5 metres wide and of course was made up of reflective walls and windows of buildings. As occurs in relatively small rooms, a number of resonant modes existed, like the ones described in chapter 9. Unhappily, the blade-pass note of the fans, 75 Hz, was very close to the frequency of one of the resonances in the long dimension of the light well. This had two effects: firstly a standing wave was set up, greatly amplifying the 75 Hz note at points where antinodes occurred (figure 4). The standing wave, though, did not have quite the same frequency as the fan note, with the result that there was a slow beat between the two as the two sounds slowly came into and out of phase with one another. This added to the annoyance value of an already irritating noise.

The fitting of absorption silencers to condensers is no easy thing. A circular, lined silencer with a central absorbent pod or splitter could easily be perched on the top, but as air is drawn in all around the perimeter at the base of each unit, silencing that would require an elaborate, specially designed silencer.

Although the existence of the standing wave was making the problem especially acute, it was also a blessing, because if it could be broken up there promised a big improvement. So, in order to do just this, a 3-metre-high screen was built right across the narrow dimension of the light well. This had as a basis an impervious membrane of 20 s.w.g. mild-steel sheet, faced on both sides with 50mm of rock wool, retained with perforated metal. It was carefully sited so that it divided the well into two lengths, which did not bear a simple arithmetical relationship to each other.

This had three effects: first of all the natural frequencies of the two sections of light well were raised to a level well away from those of the fans; secondly the absorbent faces of the screen rendered any standing waves in that direction insignificant in magnitude; and thirdly, the condensers were screened from the office window, so that the only sound which arrived at the office had been diffracted over the top of the screen. Allowing for the other reflecting walls, it was possible to calculate the extent of this.

As a result, the condenser noise was reduced to below the background noise level, which was shown in figure 59, a reduction of over 16 decibels in the 63 Hz octave band. This would probably not have been possible merely by trying to fit both intake and outlet silencers to the condensers, as absorption silencers are notoriously poor at low frequency.

In the country, of course, people are naturally very sensitive to noise and often their complaints about noise from factories would never be assuaged without demolishing the whole factory. This is because quite often complaining about noise is a way of working out a basic resentment of the presence of a factory. However, a case occurred recently where there really was cause for complaint. A chemical plant was built on a site in an area scheduled for light industrial development.

Unfortunately the planning authorities had seen fit to grant permission for a bungalow to be built on a plot of land which was to turn out to be just outside the boundaries of the chemical plant.

The unfortunate occupant of the bungalow was therefore assailed with the most extreme variety of noises in the industrial repertoire throughout the whole day and night. To start with, there were two large reciprocating compressors which gave him a hefty 88 dB outside his bungalow, some of it at the fundamental frequency of the compressors, below the audible range but capable of vibrating things from window panes to roofs, but plenty in the 31·5 and 63 Hz octave bands. He then got 58 decibels at 1 KHz from some propeller fans ventilating the compressor house, capped by the same level at 4 KHz, radiating from the long stretches of pipework handling gas under pressure and incorporating several globe valves.

As another problem, the noise level in and around the plant was in the hearing damage range. Although it was not closed in, there were enough reflections to set up a semi-reverberant field so that it was not possible to locate the high-frequency noise source just by listening. Once again, vibration measurements in various surfaces, at first made aurally, using a type of stethoscope known as a tectascope, then with an accelerometer and sound-level meter, nailed it down to a section of pipe in the neighbourhood of an exhaust valve and extending up to an exhaust stack. The stack had had fitted to it a standard type of silencer used in these plants, consisting of a cone-shaped chamber packed full of stainless-steel rings, making a type of high-frequency reactive-cum-dissipative silencer.

Sound-insulating lagging to the valve and pipework knocked the noise level down by 19 dB at 4 KHz, but a problem remained owing to the fact that the piping was rigidly clamped to the structure, and a pure tone, resulting from side-to-side oscillation of gas in the pipe, was getting into the structure and being radiated from surfaces such as chequer plating. The support brackets themselves were potent radiators. It would have been acoustically feasible to put an absorption silencer downstream of the valve, obviating the need for much of the lagging, and this would have coped with the problem of radiation from the structure. However, the engineers concerned did not want to have to put in an absorption silencer, and so it was necessary to install resilient supports for the whole length of pipe.

The compressor noise problem was purely one of intake noise. The pulsing of air into the large slow-speed compressors had harmonics well into the audible range. For reasons of economics, the company, who had their own rolling and welding facilities, decided not to opt for a standard proprietary intake silencer, and a special reactive unit was designed and made. This consisted of a large cylinder divided into two expansion chamber resonators between which a side-branch resonator was formed in an annular cavity round a central perforated pipe, which also extended, unperforated, halfway into the expansion chamber at each end.

The result was that the noise fell to inaudibility at the bungalow. The remaining problem of the propeller fans ventilating the compressor house was then a comparatively simple matter of fitting circular absorption silencers, lined with 50-mm-thick mineral wool, and with a cylindrical sound-absorbent pod in the centre. A small drop in air volume resulted because of the resistance to airflow imposed by the silencers.

In all these cases, wrong diagnosis could have resulted in a great waste of money. The cyclones could have been given damping, not sound-insulating lagging, or, worse still, they might not have been shown up as the culprit and a whole lot of treatment given to the driers or some other part of the plant. In the light well, ignorance of the effect of the standing wave could have meant a most difficult and expensive exercise of trying to duct the intakes to the condensers and incorporate silencers as well as fitting discharge silencers, and still being lucky to get 10 dB reduction. In the chemical plant, the owners were all set to fit another silencer at the base of the exhaust stack, although the problem was being caused much further back. With compressors, one frequently comes across cases where people have tried to use sound absorption, but at such low frequencies it is virtually useless unless an absorptive splitter silencer of vast dimensions can be constructed. As there is almost no middle- and high-frequency noise to worry about at the intakes to these compressors, that sort of splitter silencer would be a great waste.

Do-it-yourself silencing in factories often runs people into trouble. A large motor-car manufacturer tried to cope with a noisy dust collector in the foundry by spending £4,000 on a three-sided acoustic screen. It would have had a little effect in one direction if it had been a metre higher, but as it was not and was constructed from office partitioning with not a hint of sound absorption on either surface, on balance it made matters slightly worse because of a build-up of reflections round the machine.

On a much smaller scale I came across a small electric pump in a domestic garage below a flat. Both structure-borne noise and airborne noise were causing trouble upstairs, and in an effort to cure things a timber box lined with fibreboard had been constructed round the pump. The trouble was that the sides of the box were in contact with the pump and improved its radiation efficiency, so that the noise level in the 125 Hz octave band went up by 2 dB. In a well-meaning gesture, the pump had also been placed on a little rubber heel mat. Of course, not only was the resilience of such a thin layer of rubber nothing like enough, but all the pipework was still solidly connecting the pump to the floor and the walls.

Sometimes you can make recommendations until you are blue in the face and still get no joy because the fitters of the installation are completely unfamiliar with noise-control measures, probably do not 'hold with them' anyway and certainly do not understand them. A ventilating contractor installed a ventilation system in a small office, with the plant in a void above the ceiling. The ceiling was made up of glass panels, and not only was it necessary to fit an absorbent splitter system in the ducting, but absolutely essential to have complete mechanical isolation between the fan and the structure. Acoustic treatment on this job was an afterthought, and a method of remounting the baseplate of the fan and motor was designed. Unhappily the fitters who carried out the modifications put in the anti-vibration mountings, but left the baseplate still bolted to the beams!

Another frustrating experience occurred on a ventilating job when for various reasons it was necessary to apply acoustic lining in the form of 50-mm-thick foam with a special backing layer of damping sheet to a length of sheet-metal ducting. Unfortunately the installers excelled themselves in common sense and came to the laudable conclusion that there would be better airflow if the foam was installed with the damping sheeting facing inside the duct instead of being stuck to the sheet metal. Some poor fellow had to crawl along the duct later picking off the damping layer with his fingernails!

I once went with a heavy heart to a factory whose chief engineer told me that he had had several acoustical consultants in and all their proposals had been useless. It turned out that the noise was due to such a multiplicity of sources in a reflective building that rather than go for clumsy enclosures for each of the many complex machines the different consultants had all recommended installing sound-absorbent cladding to the ceiling, and/or filling in the roof trusses with acoustic boards. Alas, he decided on a pilot scheme, treated one roof truss, and wondered why it made no difference. Of course the meagre bit of extra absorption he got from this small area nowhere approached the doubling of absorption required before you get a 3 dB drop in reverberant sound.

I have already referred in this book to the case of the pegboarding, which unfortunately is quite true. People really like the idea of small holes sucking up noise, and on two occasions I have seen machines enclosed with nothing but pegboard, and sometimes for good measure a sound-absorbent backing. Neither, of course, had any worthwhile effect, but they did cost money and highlight the real need for more dissemination of the science of acoustics.

15   The noisy future

What has the future in store for the world of noise? Is it going to get noisier, or quieter? What new techniques will there be, what new methods for controlling noise? In many spheres it is going to be a case of having to run hard in order to stand still. With the likelihood of larger and heavier jumbo lorries, jumbo jets and jumbo everything, if we are not careful we will have jumbo noise to go with them.

We are always having to remember that quiet costs money. This means that no machine, vehicle or aircraft engine manufacturer is going to pursue quietness simply for philanthropic reasons. Even now the strides which have been made in the designs of quieter turbo-fan engines will probably be of no benefit to ground dwellers. Give an airline operator a quieter aeroplane, and his pilots will use more power after take-off and still keep within the limits of airport noise-abatement regulations. You will not hear any difference on the ground.

A machine manufacturer will try to make a quieter product only if he is forced to either by legislation or because customers want quiet machines and will choose a rival product for its lower noise level. A machine buyer will not choose to pay for a quiet machine unless there is good reason to. Those good reasons exist, but they have got to be dragged out into the light of day. Even then, it will still take legislation before the majority of those responsible will do anything. I know of cases where management are fully and completely aware that the noise from their manufacturing process is severely damaging workpeople's hearing, have the means of doing something about it at their fingertips and yet do nothing.

More legislation has got to come about; existing legislation must be tightened up. We can probably expect a new Factories

Act which will include limits for noise exposure. Nevertheless the real problem is basic: lack of knowledge. The effects of noise are not fully understood and some people refuse to believe that it can do your hearing any harm. Perhaps more important, though, is the fact that design engineers, with some notable exceptions, even if they do have some knowledge of acoustics, forget all about it. In every one of the case histories in the previous chapter the acoustic treatment was an afterthought because *nobody realized* that the machines would create a noise problem. There should be a poster on the wall of every design office asking 'What about noise?' It will cost a few pounds perhaps if it is necessary to summon an acoustician to look at the drawings, but how much happier will everyone be when the product starts up quietly in the first place. How much cheaper it is to build quietness in than to take everything to bits later and try to patch things up.

Acoustics is not a difficult subject, and, once the basic facts are understood, applied common sense will get you a very long way. There are many skilled and experienced specialists now both in the universities and in consulting organizations who can readily cooperate with designers. There are many manufacturers of acoustic equipment and materials who offer not only excellent products but a wealth of know-how.

Already consumer demand is beginning to become very noise-conscious. The word 'noise' appears more and more in motor-car advertisements and industrial suppliers are being asked to quote noise levels as well as other technical data. Unfortunately they all too often come back with a bald statement that 'the machine noise is 76 dB' and this has very little meaning unless qualified. They should at least quote the testing procedure used; if a single figure is given it must be in dBA, using the weighting network of a sound-level meter which reduces its response to low and very high frequencies so as more nearly to match the response of the human ear. If our plain 76 dB referred to noise which was predominant at 60 Hz it would be about the same loudness as another noise of only 36 dB which centred around 500 Hz. Measured in dBA these two figures would be 50 dBA and 53 dBA respectively – much more realistic.

Of greater use, of course, are octave or even narrower band

levels, provided that the test procedure is also quoted, but the
equipment necessary to do this can be expensive. However, for a
standard product, testing can be carried out by one of the acoustical laboratories and detailed results made available. The actual
description of testing procedures is beyond the scope of this book.
British Standard 4196: 1967 gives a guide to the selection of
methods of measuring noise emitted by machinery, but in addition
many research associations and similar bodies in various industries
have evolved standards.

What technical developments can we expect in the future?
What will the noise from machines be like? The answer is that
some will be noisier, some quieter and some the same. The aero-
engine industry will probably just about keep pace with the rising
size of jet engines and prevent large increases in noise. Vertical
take-off and landing will drastically alter the noise-contours
around airports, increasing the noise in the immediate neighbourhood, but as a greater altitude can be reached in a shorter distance,
inhabitants in the medium field will benefit, those in the far field
will probably find little difference.

Diesel engines are in grave danger of becoming much noisier
as casting methods improve and unstressed parts of the engine
are made thinner and thinner. When the walls of a crankcase
become as thin as 5mm, noise radiation starts to go up at a greatly
increased rate. Much of the necessary research on engine
noise has already been done, and it has long been possible to
silence exhausts very satisfactorily. It is simply a matter of cost,
and unless consumer demand or the law forces engine manufacturers to adopt silencing measures, it would be suicidal for them
to add costs when they can sell as many noisy engines as quiet
ones.

In fact all through industry the trend is to lightness. Lightweight
building methods mean lower mass and less sound insulation, and
alternative sophisticated composite partitions of lightweight are
expensive. It must be fully realized that thermal and acoustic
insulation are not the same thing. Research is constantly in
progress in the building industry and there is no shortage of
know-how. We shall see 'soundproofing' as a package available
in a house rather like central heating for perhaps a 10 per cent

increase in price. This will include sealed doors, double-glazed windows, of the acoustic type, not the plain thermal type, acoustically treated ventilation systems, lined roofs and perhaps sound-absorbent ceilings in the rooms. Partitions of two or more resiliently mounted skins can be used where heavy masonry is undesirable.

We can certainly expect some improvement in some corners from the use of new or different materials. Plastics have much better inherent damping than steel, and we have already seen the use of carbon fibres in fan blades of aero-engines. We shall see more and more plastic motor cars, and it is partly because of the money tied up in metal working plant that we do not see more now. Of course plastic has less mass than steel, and if external damping is applied to the latter it becomes a good insulator. However, in motor cars, for instance, the only place where sound insulation is more important than damping is in the bulkhead. It will be perfectly possible either to use externally applied loaded plastic sheets, as are seen now in some good-quality saloons, or possibly to load the basic plastic in critical areas. The main problem will be a lowering of insulation against external noise, and it might be rather unpleasant for drivers in heavy traffic with heavy diesel lorries leering at them on both sides.

One of the biggest changes in the prevailing noise climate will probably come with a change of prime movers. The days of the reciprocating internal combustion engine are numbered, even if that number is rather large. When electric propulsion gets a real foothold it will bring about as big a reduction in the noise in the street as we shall probably ever see, to say nothing of the reduction of other forms of pollution. All hangs on the battery men at the moment, and until it is possible to carry around enough power to go more than about 60 kilometres on a battery-full and travel at more than 70 k.p.h., the appeal of the electric vehicle is limited. Nevertheless, even now a 60 kilometre range and 70 k.p.h. maximum speed would be perfectly adequate for purely town use. Once through traffic can be eliminated, we may see internal combustion engines banned from city centres except when heavy loads are to be transported to or from a place in the city. What pleasant places cities would be with little noise, no exhaust and parking for all! Electric vehicles are not completely noiseless, and motor-hum would be there together with transmission noise, but a bit of acoustical engineering on the part of the designers would make it insignificant.

Another source of motive power, of course, is the gas turbine. This, at present, has some engineering drawbacks but these will no doubt be overcome. It does not necessarily produce a lower basic sound level than a conventional engine, but it is sound which is much easier to cope with. There is no jet, and gone are all the low-frequency pulsing, throbbing, pounding components of diesel noise. What noise there is is of a frequency content which is much easier to deal with. Almost all silencing methods are painfully inefficient at low frequency.

In the air, it will be a very long time before we see, or rather hear, anything other than a variation of the turbo-fan engine. One ventures into the realms of science fiction in trying to think up the next generation of aero-engines. I know of one group who are busily working now on what they call an 'anti-gravity device', but if they get anywhere they will be on to something beyond the scope of modern physics. It would be lovely to be able to rectify the alternating force created by a pair of contra-rotating eccentric bodies, whether lumps of lead or electrons. You can do it in a very inefficient manner by placing them in a vehicle which prefers to travel forwards rather than backwards, but I cannot see it being used for air or space travel. As science certainly has not reached its limit, we may yet see an anti-gravity device, and assuming that it is quiet (what a blow if it turns out to be noisy!) then one more major noise problem can be overcome. I am not very hopeful.

A little nearer to reality, but still spectacularly impractical for anything but interplanetary travel, is the electric engine. Modern air and space travel is achieved because of the reaction against the acceleration of mass. With jets, it is air and the products of fuel combustion, with space rockets there is no air, just hot gas. An electric engine relies on the minute mass possessed by electrons, which are accelerated away from the engine, and the reaction to this produces about the same power as a strong handshake.

In space, of course, the longevity of an engine is of more value than its power output because after a few years at constant small acceleration in a vacuum you get quite a move on, much more so than with a few hours violent blast. Electric engines now, though, are completely unsuited to air or ground travel because of their tiny power output. I do not think the motoring magazines would rejoice in reporting 0–100 k.p.h. acceleration times in the region of 120,000 seconds!

A possible newcomer in the field of prime movers may involve harnessing the energy possessed by expanding and contracting *solids*, as opposed to gases. A reciprocating engine working on this principle could be a great deal quieter than a gas engine. For one thing, it would run at a speed so low that the counterpart to combustion noise would be inaudible, and very poorly radiated. If an engine like this also incorporated magnetic bearings, which are now still in their infancy, it might almost warrant the description of 'silent engine'. Magnetic bearings themselves, and similar devices, could be put to very profitable use on existing machines, and of course acoustics is not the only profession in which there is interest in them.

There will certainly be some new materials which combine conventional acoustical physics into a practical, economical and highly efficient form. Highly damped, low-stiffness and multiple-skin materials have many uses, and improvements in damping methods will mean that much needed stiffness can be retained. A typical example is that of highly damped steel sheet – a light-gauge sheet bonded to the main sheet with resin so that a constrained damping layer is formed by the resin. This is already being made in small quantities by the British Steel Corporation, and although, as we saw earlier in the book there are difficulties associated with working this material, they will eventually be overcome. It will always cost considerably more than ordinary steel sheet. None of this, though, comes under the heading of a real breakthrough.

But there is one path which is virtually untrodden. You may have noticed what a passive business silencing is. We are always relying on mass reactances or reactions of one sort or another (in the physical sense of the word, not the chemical sense). The incident sound wave either supplies the energy necessary to thwart it, or relies on something as simple as inertia.

If there are going to be any breakthroughs they will probably come in the field of artificial reaction to sound waves. There are two or three ways in which this might be done, and some have been tried on a limited scale. The first possibility is very attractive, and involves the generation of 'anti-sound', which will combine destructively with sound and annihilate it.

Anti-matter is made of atoms whose particles have the opposite charge to those of matter, and in isolation is indistinguishable from matter. Only when the two meet do they annihilate each other. The pressure fluctuations of anti-sound would have the opposite sign to sound. In isolation, anti-sound would be no different from sound, but when the two met they would cancel each other out.

It sounds wonderful, but there is a catch. It is easy to generate anti-sound simply by sensing the sound with a microphone and reproducing it with an inverted waveform. What is nearly impossible to do is to produce an anti-sound wave front exactly the same shape as that of the sound so that the equal and opposite pressure fluctuations occur in the right place and the right time. Take a simple spherical source, for instance, like the pulsating balloon from chapter 2. To reproduce an identical spherical wave front of anti-sound to annihilate the sound from the balloon by conventional means would require that the source of anti-sound should be located at the centre of the balloon. This might be possible, but not many of us suffer sleepless nights because of the noise of pulsating balloons, and if instead you take a more elaborate machine the radiating surface of anti-sound would have to be the same shape and occupy the same space as the machine.

Partial success is possible if one abandons the idea of trying to annihilate the sound radiating in all directions from the source and is merely content with creating a small patch of quiet. This occurs on a passive scale of course when interference between an incident and a reflected wave results in a standing wave (figure 6). Here the reflected wave is the anti-sound, and at a series of points, for simple waveforms, the two waves always cancel each other out, causing little or no sound at all at those points. Unfortunately those points are literally 'points' and even if you can manoeuvre your ear into exactly the right position it will still hear sound on either side of the point. Anyway the effect will be spoiled because the other ear will be out of the silent point except for simple tones whose wavelength is the right length to enable you to have one ear in one silent spot and the other in another. Bearing in mind that a reduction of 20 dB is a drop in sound pressure of 90 per cent, the cancellation between the two waves has got to be pretty good to have a decent effect.

It is possible greatly to extend the area of silent spots by creating a wave front of anti-sound much more nearly the shape of the sound. With incident and reflected sound the silent spot is so small because with spherical waves and a flat reflecting surface the wave fronts could hardly be more different, as the curvature is in the opposite direction. In addition, the effect is only worthwhile for a simple waveform. If an anti-sound wave front is generated which approximates to that of the sound, the silent spot is greatly extended in area, and as the anti-sound is artificial there is no need to restrict the experiment to simple waveforms. Any waveform can be synthesized, and, if necessary, time-delay mechanisms can be built in to ensure the right phase relationship.

One of the other means of introducing artificial reaction might possibly be in the field of sound-insulating partitions. If one constructed a partition and established its modes of vibration to different frequencies, electromagnetic exciters could be fitted at antinodes, or points of maximum displacement. These would often unfortunately be many in number and in different places for different frequencies. Then by means of a network of microphones on guard a short distance ahead of the panel, warning of the waveform of an advancing wave front could be fed to a time-delay mechanism and subsequently fed to the exciters so as to produce equal and opposite reactive stresses on the panel to the force applied by the incident sound. This method, of course, has severe limitations, not least of which at present would be the sheer complexity and cost of the necessary electronic equipment. It would require a major research programme even to get a model working in a laboratory. In the first instance the panel and exciters would have to be mounted in an immensely stiff frame, or the performance would be limited by factors like the mass of the exciters, and except at high frequencies this would not be worth having. At high frequencies and oblique angles of incidence, the panel would be moving in several different modes at the same time, partly overcoming the latter problem, but greatly increasing the number of exciters required. There would only be hope for this sort of thing if the electronics could be made anything approaching simple, and even then the cost could seldom be justified.

Electronics are used, though, in the field of concert-hall acoustics. A system known as 'assisted resonance' is used when the building design is such that it is difficult to get a long enough reverberation time. In principle this is done by the electronic reproduction of sound via loudspeakers in strategic positions at selective frequencies, but the effect is very different from that of a basic amplifier system. Another method uses a time-delay system so that the original sound is repeated over and over again with diminishing amplitude, simulating the effects of reflections of sound from wall to wall. This can be useful in small recording studios with highly absorbent walls where it is required to simulate a large hall or similar building.

The problems of artificial reaction and anti-sound would be a little less acute in such fields as exhaust silencing, where the dimensions of the 'waveguide' are such that we are dealing with straightforward plane waves and do not have to worry about wave fronts. The problems would be more ones of temperature, corrosion and general rough treatment of delicate components. Again the cost would be terrific, and conventional silencers can do a very good job as they are.

I am often asked at lectures why nobody has thought of using a vacuum as a sound insulator, as it was so efficient in the fifth-form physics class, using a bell in a jar with the air pumped out of it. It is quite true, of course, that sound cannot exist in a vacuum, but not only is a vacuum about the most expensive way there is of getting nothing for your money (although I can think of some strong competitors), but it also requires containing, or rather excluding.

A simple sandwich panel with a vacuum in the middle would

not be startlingly effective because of the sound which would bridge the gap across the mechanism which holds the panels apart, and even if you used such a resilient means of doing this that you got 90 per cent isolation, you still would not improve on conventional partitions. I do not see much of a future for 'vacuum-packed' panels, although I suppose if you had a machine which would stand up to it, was supported from the ground electro-magnetically and enclosed in a strong enough structure to allow the space to be evacuated, you would not have much trouble with noise. You would have other troubles though. What use is a machine in that sort of isolation?

However, no scientific miracle is necessary to secure a steady and large improvement in noise today. Indeed, a reversion to some nineteenth-century ideas would promise a substantial reduction in noise; the steam engine and the Stirling engine are both heading for the limelight again, with the benefit of modern technology, and because they neither of them rely on explosions in their cylinders, their re-emergence promises to be gratifyingly peaceful. We know enough already to enable us to reduce the noise from many sources. 'Where there is a will, there's a way.' We have the way without the will. If we want quiet, we can have it, if we pay for it. Whether it means paying for airports with flight paths over the sea, or quiet running engines and machines, it can be done.

Glossary of acoustical terms

ABSORPTION COEFFICIENT (α): If a surface is exposed to a sound field, 'α' is the ratio of the sound energy absorbed by the surface to the total sound energy which strikes it; if it absorbs 60 per cent of the incident energy the absorption coefficient is 0·6.

ACOUSTIC: Having properties or characteristics affecting or connected with sound; 'acoustic tiles', but not 'acoustic engineer' (unless you are referring to his sound absorption coefficient).

ACOUSTICAL: Relating to *acoustics* – 'acoustical consultant'.

AMBIENT NOISE: The background noise or prevailing general noise in an area, sometimes in the absence of a noise of particular interest.

AMPLITUDE: The maximum value, the peak.

ANECHOIC: Almost totally sound-absorbent at a very wide range of frequencies. An anechoic chamber gives almost *free field* conditions.

ANGULAR FREQUENCY (ω): For mathematical reasons it is often simpler to multiply the *frequency* in *Hertz* by 2π, which is the equivalent of an angle of 360° in radians. Angular frequency is then the frequency expressed in radians per second.

ANTINODE: A point, line or surface where the vibration *amplitude* is at its maximum (see also *node*).

AUDIO FREQUENCY: A frequency within the audible range of about 20 Hz to 20,000 Hz.

AUDIOGRAM: A graph, usually automatically plotted by an audiometer, showing a subject's hearing response or loss as a function of frequency. A separate graph is usually given for each ear.

AUDIOMETER: A machine for producing an audiogram normally automatically feeding calibrated pure tones to each earphone and plotting the levels at which the subject signals he can or cannot hear them.

BAND: A segment of the frequency *spectrum*, i.e. an octave, half octave, third octave.

BEAT: Periodic increase and decrease of amplitude resulting from the *superposition* of two tones of different frequencies f_1 and f_2. The beat frequency is equal to $f_1 - f_2$.

BEL: Ten *decibels* (not normally used).

BINAURAL: Utilizing the facility of hearing in two ears, for instance for range or direction finding. Also, simulation of this effect electronically.

BONE CONDUCTION: The means by which sound can reach the inner ear and be heard without travelling via the air in the ear canal or meatus.

CHARACTERISTIC IMPEDANCE (ρc): A measure of the qualities possessed by a substance carrying sound waves which indicates the ratio of the *effective sound pressure* at a point to the effective *particle velocity*. It is equal to the product of the density 'ρ' and the speed of sound 'c' in the substance, sometimes given the name 'Rayl'. For air the characteristic impedance is 408 Rayls (in m.k.s. units) at 20° C and a pressure of 1 bar.

COCKTAIL PARTY EFFECT: The faculty of 'locking-on' to one voice amid a hubbub of other voices.

COINCIDENCE: When the length of a bending wave in a panel coincides with the length of an incident sound wave at the angle at which it strikes the panel. There is a frequency below which coincidence cannot occur called the *critical frequency*.

CONTINUOUS SPECTRUM: A frequency analysis having components continuously ranged over the spectrum.

COUPLED MODES: Modes of vibration which mutually influence one another.

CRITICAL FREQUENCY: The lowest frequency at which *coincidence* can occur in a panel, and above which the sound insulation is reduced. The lower the stiffness and the thinner the panel, the higher the critical frequency. Appendix 1 shows some typical values.

CYCLE PER SECOND: See *frequency*.

DAMAGE RISK CRITERION: The noise level as a function of frequency and factors such as waveform (i.e. pure tone or *random*) and intermittency above which permanent hearing loss greater than a specified amount is likely to be sustained by a person subjected to it.

DAMPING: Removal of energy from an oscillating system or particle by means of friction or viscous forces. The energy is converted into heat.

DECIBEL (dB) (one tenth of a Bel): A means of denoting the ratio of two quantities when the range of values is very great. A Bel can be described as the number of tenfold increases the lower quantity must be given to equal the higher, i.e. $\log_{10} \dfrac{p_1}{p_2}$;   to obtain the answer in decibels, multiply by 10. *Sound pressure level* is the commonest quantity expressed in decibels, in which case the lower quantity is usually 2×10^{-5} N/m², known as the reference pressure. The sound pressure level in dB is more or less numerically equal to the sound *intensity level*, when the reference is 10^{-12} watts/m². 'dBA' are sound pressure levels in decibels measured on a sound *level* meter incorporating an 'A' weighting network which reduces its response to low- and very high-frequency sound in order more nearly to simulate the response of the human ear and to give results which give some indication of the loudness, annoyance value or acceptability of a sound; dBA often work out 10 units higher than the equivalent *Noise Rating Number* of the noise.

DIFFRACTION: The diversion of the direction of travel of a wave other than by reflection or *refraction*. A typical example is the diffraction of sound into the *sound shadow* formed by an acoustic screen.

DIFFUSE FIELD: A sound *field* in all parts of which the sound pressure level is the same, and the sound waves are equally likely to be travelling in any direction.

DISTURBANCE: See *excitation*.

DOPPLER EFFECT, DOPPLER SHIFT: The change of frequency of sound as observed at a point in relative motion with the source and/or the medium carrying the sound. When a vehicle on a road sounding its horn passes a stationary observer, in the interval between one oscillation of the horn diaphragm and the next the vehicle and the horn itself have to a small extent kept pace with the first sound wave, so that the second is created closer to it than if the vehicle had been stationary. A shorter *wavelength* means higher *frequency*.

ECHO: Reflected sound which arrives a long enough time after its direct equivalent for it to be heard as a separate sensation.

EFFECTIVE: When a quantity is described as effective, such as in the phrase 'effective sound pressure', it means the *root-mean-square value*.

EXCITATION: A forced variation in pressure, position or similar quantity.

FIELD: A region of acoustical interest.

FORCED OSCILLATION OR VIBRATION: *Oscillation* or vibration maintained by an applied fluctuating energy supply (see also *natural frequency*).

FREE FIELD: A region in which no significant reflections of sound occur.

FREE PROGRESSIVE WAVE: A theoretical wave propagated in an infinite medium.

FREQUENCY: The number of times a vibrating system or particle completes a repetitive cycle of movement in a period of one second, expressed in Hertz or 'cycles per second'. Non-periodic waves can also be defined in terms of frequency, in which case the rates of rise and fall of pressure for a given amplitude govern the frequencies in the wave.

FUNCTION: A quantity which varies as a result of variations of another quantity.

FUNDAMENTAL FREQUENCY: The *frequency* with which a *periodic function* reproduces itself, incorrectly described as the lowest frequency of a complex periodic wave, sometimes called the first *harmonic* (see also *sub-harmonic*).

GAUSSIAN DISTRIBUTION (or Normal Distribution): A term used in statistics to describe the extent and frequency of deviations or errors. The outstanding characteristics are a tendency to a maximum number of occurrences at or near the centre or mean point, the progressive decrease of frequency of occurrence with distance from the centre, and the symmetry of distribution on either side of the centre. In respect of *random noise*, each fluctuation of amplitude is an occurrence, whether above or below the mean level; the peak value of each fluctuation is the error and the distribution of errors with time is Gaussian.

GRADIENT: A variation of the local speed of sound with height above ground or other measure of distance causing *refraction* of sound. It is most commonly caused by rising or falling temperature with altitude or by differences in wind speed.

HARMONIC: A *sinusoidal* (pure-tone) component in a complex periodic wave, of frequency which is an integral (whole number) multiple of the *fundamental frequency* of the wave. If a component (an 'overtone') in a sound has a frequency twice that of the *fundamental* it is called the second harmonic.

HEARING LOSS: The amount in decibels for a specified ear and frequency by which the *threshold of audibility* for that ear exceeds the normal threshold.

HERTZ (Hz): See *frequency*.

IMPEDANCE: A measure of the complex ratio of force (or pressure) to velocity; see also *characteristic impedance*.

INFRA-SONIC: Of *frequency* below the *audio frequency* range.

INTENSITY: The rate of energy flow per unit area transmitted as a sound wave, usually expressed as a decibel ratio of watts/m². For *plane* or spherical *free progressive waves* it is equal to $\dfrac{p^2}{\rho c}$ where p = sound pressure, ρc = the *characteristic impedance* of the medium.

LEVEL: The value of a quantity in decibels.

LOUDNESS: The judgement of intensity of a sound by a human being, or just the ear. It is dependent on sound pressure and frequency. Over much of the range, a threefold increase in sound pressure is considered a doubling of loudness. This is a change of just under 10 dB.

MASKING: The raising of the *threshold of audibility* of a sound by the presence of another sound. It is most marked when the masked sound is of higher frequency than the masking sound.

MEAN FREE PATH: The average distance sound travels between successive reflections in a room.

MEDIUM: A substance carrying a sound wave.

NATURAL FREQUENCY: The frequency at which a system oscillates freely after suitable excitation.

NODE: A point, line or surface where a wave has zero *amplitude*.

NOISE: Sound unwanted by the listener, meaningless sound, *random* sound.

NOISE CRITERIA: Sets of curves relating levels of sound in octave bands to speech interference and acceptability for particular applications, usually types of office.

NOISE LEVEL: Sound level.

NOISE RATING CURVES OR NUMBERS: Sets of curves relating levels of sound in octave bands to acceptability for particular applications, from factory noise to noise in homes. When an octave-band analysis is plotted on a graph of NR curves, the number of the curve which is reached by the level in one or more bands is the Noise Rating Number (NRN) of the noise. There is also an arithmetical method of achieving the same result. In general application dBA tend to be more popular and useful than NRN.

NOISE REDUCTION COEFFICIENT (NRC): The average of the *absorption coefficients* of a surface or material at 250 Hz, 500 Hz, 1KHz and 2KHz. A quick guide to the usefulness of acoustic tiles.

NOY: A unit of noisiness related to *perceived noise level* in PNdB.

OCTAVE: The interval between two sounds one of which has a frequency twice that of another.

OCTAVE BAND: See *Band*.

OSCILLATION: Variation in the magnitude of a quantity above and below a certain level over a period of time (or distance).

PARTICLE: A theoretical infinitesimally small part of a substance or medium.

PARTICLE VELOCITY: When a sound wave passes a point in a medium the particle at that point must oscillate in order for the sound wave to be transmitted. The velocity of the particle either at a particular instant, or its maximum, or its *root-mean-square* or *effective* value is of interest, and for materials of a given density and a given *effective* sound pressure, the *effective* particle velocity varies with the characteristic impedance.

PEAK SOUND PRESSURE LEVEL: The value in decibels of the maximum sound pressure, as opposed to the *root-mean-square* or *effective* sound pressure.

PERCEIVED NOISE LEVEL (PNdB): The sound pressure level of between one third of an octave and an octave of *random* noise at 1,000 Hz which is considered by 'normal' people to be equally noisy to the sound of interest. PNdB = $40 + 10 \log_2$ Noy.

PERIOD: See *periodic*.

PERIODIC: Repeating in an identical form after a constant and repetitive period of time. The classic example is *sine wave*.

PHASE: A measure of whether a sound or other *periodic* function is 'in step' or 'out of step'. It is measured as an angle either in degrees or, better, in radians ($360° = 2\pi$ radians) and if, for instance, one *sine wave* lags behind another so that it is always at its minimum when the other is at its maximum it is π radians or $180°$ out of phase. (See also *angular frequency*.)

PHON: A unit of loudness level (see *sone*).

PITCH: An aural assessment of sounds so that they can be ordered in a scale from low to high. It is primarily dependent on *frequency*, but also on sound pressure and waveform.

PLANE WAVE: A wave in which the *wave fronts* are parallel with one another at right angles to the direction of propagation.

PNdB: See *perceived noise level*.

PRESBYCOUSIS OR PRESBYACUSIS: Hearing loss due to advancing age, usually at high frequency.

PURE TONE: A sound whose waveform is *sinusoidal*.

RANDOM NOISE: Strictly speaking, a fluctuating quantity (not necessarily sound – it may be electronic) whose amplitude distribution with time is *Gaussian*. Generally, noise due to random pressure or other fluctuations resulting in a *continuous spectrum*.

RAYL: See *characteristic impedance*.

REFRACTION: The bending of sound by passage from one medium to another or in a *gradient*.

RESONANCE: When a system is vibrating as a result of a forced *excitation* at a certain frequency, if the *amplitude* of vibration diminishes as a result of raising or lowering the frequency of the exciting force then the system is in resonance.

RESONANT: Capable of being excited into resonance.

RESONANT FREQUENCY: A frequency at which *resonance* occurs.

REVERBERATION: Sound at a point which builds up owing to multiple reflections from surrounding surfaces. It will persist after the source has stopped emitting sound.

REVERBERATION TIME: The time it takes for reverberant sound of a given frequency to decay by 60 dB after the source is cut off. It is usual to measure the first 30 dB decay and extrapolate the rest.

ROOT-MEAN-SQUARE (RMS) VALUE: The effective value of a fluctuating quantity. The values of a quantity are squared, averaged and then the square root is extracted. The *peak sound pressure* of a sine wave is equal to the effective sound pressure multiplied by $\sqrt{2}$. The effective sound pressure level is the best measure of ordinary sound, but the peak level is necessary for assessment of impulsive noises.

SINE WAVE: A *wave* which varies with time or distance as the trigonometric function, the sine. It is the purest type of wave, and is sometimes called a pure tone. If a point on the circumference of a rotating circle is projected on to a straight line, its displacement will vary as the sine of the angle subtended by the point and the top of the circle.

SINUSOIDAL: Varying as the sine of an angle.

SONE: A unit of loudness designed to give scale numbers roughly proportional to the loudness. $Phon = 40 + \log_2 Sone$.

SOUND: Wave motion in an elastic *medium*, or the sensation of hearing this may produce.

SOUND INTENSITY: See *intensity*.

SOUND POWER LEVEL: The total energy per second emitted by the source as sound expressed in *decibels*, normally re 10^{-12} watts.

SOUND PRESSURE LEVEL: The *effective* sound pressure, or *root-mean-square* values of the pressure fluctuations above and below atmospheric pressure caused by the passage of a sound wave, expressed in *decibels* re 2×10^{-5} N/m².

SOUND SHADOW: The acoustical equivalent to a light shadow, usually partially filled in as a result of *diffraction*.

SPECTRUM: A group of continuous frequencies rising from low to high (see also *audio frequency*).

SUBHARMONIC: A *harmonic* of frequency an integral number of times lower than the *fundamental frequency* in a periodic wave.

SUPERPOSITION: The arithmetical combination of two or more waves at successive instants or points.

THRESHOLD OF AUDIBILITY: The minimum sound-pressure level at which a person can hear a sound of a given frequency.

THRESHOLD OF PAIN OR FEELING: The minimum sound-pressure level at which a person starts to sense feeling or pain in the ear from sound of a given frequency.

THRESHOLD SHIFT: An alteration, either temporary or permanent, in a person's *threshold of audibility*.

TONE: A sound of definite *pitch*.

TRANSMISSION LOSS: A measure of the sound insulation of a partition, wall or panel in decibels. It is equal to ten times the logarithm of the ratio of the intensity of the incident sound wave to the intensity of the transmitted wave, or simply the arithmetical difference between the level in decibels of the incident and transmitted waves. However, when a total enclosure is involved, reverberation inside the enclosure will reduce the overall sound insulation.

WAVE: A disturbance propagated in a medium.

WAVE FRONT: A theoretical surface which is made up of points at which the *phase* of a *wave* is the same. With a sine-wave the wave front joins all points of equal amplitude and phase.

WAVELENGTH (λ): The distance between 'crests' of a sine-wave, or, more correctly, the perpendicular distance between two wavefronts in which the phases differ by one complete period. It is equal to the speed of sound divided by the frequency. Random or non-periodic sound can be defined in terms of *frequency* and therefore also wavelength.

WHITE NOISE: Noise of a statistically random nature having equal energy at every frequency over a particular *band*.

Appendices

Appendix 1. Surface density and critical frequency for various
materials

Material	Surface Density Kg/m^2	Young's Modulus (E)	Critical Frequency $\times$ Surface Density $Hz \times Kg/m^2$
Lead	11	14×10^9	600 000
Steel	8	210×10^9	98 000
Aluminium	2·7	70×10^9	32 000
Glass	2·5	41×10^9	38 000
Concrete	2·3	24×10^9	44 000
Brick	2	16×10^9	42 000
Hardboard	0·8	$2·1 \times 10^9$	30 500
Plasterboard	0·75	$1·9 \times 10^9$	29 000
Plywood	0·6	$4·3 \times 10^9$	13 200
Flaxboard	0·4	$1·2 \times 10^9$	13 200

Appendix 2. Typical transmission loss figures

	Thick-ness mm	Surface density Kg/m²	Average	\<br\>125Hz	*Transmission Loss – dB*\<br\>250Hz	\<br\>500Hz	\<br\>1KHz	\<br\>2KHz	\<br\>4KHz
Plywood	6	3	18	10	13	17	22	24	21
Glass	5	13	23	17	21	25	26	23	26
Steel	0·12	9·3	29	20	23	27	32	34	41
Brick (plastered)	145	220	45	34	36	41	51	58	60
Concrete	152	350	47	35	39	45	52	60	67
Metal-stud partition faced on both sides with double layer of gypsum wallboard	100	43	43	25	37	44	52	55	45
The same, with wood-studs	100	40	40	25	31	40	46	53	48
Plywood, glued to both sides of 25mm ×76mm timber studs	90	12	26	16	18	26	28	37	33
Proprietary modular acoustic panels	76	42	41	24	30	37	41	46	45

Appendix 3. Decibel table

Intensity Ratio	dB		Intensity Ratio	dB
100	+20		0·80	− 1
79	+19		0·63	− 2
63	+18		0·50	− 3
50	+17		0·40	− 4
40	+16		0·32	− 5
32	+15		0·25	− 6
25	+14		0·20	− 7
20	+13		0·16	− 8
16	+12		0·13	− 9
13	+11		0·10	−10
10	+10		0·08	−11
7·9	+ 9		0·063	−12
6·3	+ 8		0·050	−13
5·0	+ 7		0·040	−14
4·0	+ 6		0·032	−15
3·2	+ 5		0·025	−16
2·5	+ 4		0·020	−17
2·0	+ 3		0·016	−18
1·6	+ 2		0·013	−19
1·3	+ 1		0·010	−20
1·0	0			

Example: Add 86 dB, 83 dB and 79 dB. $1·0+0·5+0·2 = 1·7$
$= +2$, i.e. 88 dB

Appendix 4. Typical absorption coefficients

Material		Absorption Coefficient α				
	125Hz	250Hz	500Hz	1KHz	2KHz	4KHz
Air at room temperature per 1,000m³					0·06	0·2
Brickwork	0·05	0·04	0·02	0·03	0·05	0·05
Concrete	0·02	0·02	0·02	0·04	0·05	0·05
Glass windows	0·2		0·1		0·05	
Plywood or hardboard over an airspace	0·3		0·15		0·1	
Wooden floors	0·15	0·2	0·1	0·1	0·1	0·1
Thin carpet on concrete floor	0·1	0·15	0·25	0·3	0·3	0·3
Pile carpet over thick felt	0·1	0·25	0·5	0·5	0·6	0·65
76mm rock wool faced with perforated metal (solid backing)	0·4	0·7	0·75	0·9	0·85	0·75
The same 25mm thick (as acoustic tiles)	0·1	0·3	0·6	0·75	0·8	0·8
15mm fissured mineral tiles (solid backing)	0·1	0·25	0·83	0·87	0·64	0·52
12mm perforated wood-fibre tiles	0·2	0·55	0·6	0·6	0·65	0·8

Appendix 5. Attenuation of sound in acoustically lined air ducts

Thickness of lining (rock wool) on 2 sides	Width of airspace	Attenuation-decibels per 300mm					
		125Hz	250Hz	500Hz	1KHz	2KHz	4KHz
25mm	25mm	1·0	2	5	15	37	58
	50mm	0·5	1	3	10	27	26
	100mm	–	0·5	2	7	17	10
	200mm	–	0·5	1	5	7	3
	400mm	–	–	1	5	2	–
50mm	25mm	2	5	11	24	32	48
	50mm	1	3	9	19	22	28
	100mm	0·5	2	6	11	12	10
	200mm	0·5	1·5	3	7	6	2
	400mm	–	1	2	4	1	–

Appendix 6. Further Reading

Teach Yourself Acoustics, G. R. Jones, T. I. Hempstock, K. A. Mulholland, M. A. Stott. The English Universities Press.

Acoustics, Noise and Buildings, P. H. Parkin and H. R. Humphreys. Faber & Faber.

Noise and Man, William Burns. John Murray.

Noise, Final Report, Committee on the Problem of Noise. H. M. S. O.

Sound Insulation and Noise Reduction, British Standard Code of Practice, chapter 3 (C. P. 3). British Standards Institution.

Building Acoustics, B. F. Day, R. D. Ford, P. Lord. Elsevier.

Handbook of Noise Control, edited by C. M. Harris. McGraw Hill.

Noise Reduction, Leo L. Beranek. McGraw Hill.

Handbook of Noise and Vibration Control, edited by R. A. Warring. Trade and Technical Press.

Acoustics and Vibration Physics, R. W. B. Stephens and A. E. Bate. Edward Arnold.

Index

Some other Pelicans are described on the following pages.

a Pelican Original

Man and Environment
Crisis and the Strategy of Choice

Robert Arvill

What will the world look and be like tomorrow?
Must the landscape be an extension of today's spreading
deterioration? More air fouled by noise and poisoned
fumes; more water polluted by chemicals and oil slicks;
more land crushed under the sprawl of towns, super-
highways, airports, factories, pylons, and strip-mines?
Is man bound to build a stifling steel-and-concrete hell
for himself? Or can effective steps be taken now to
preserve our open-spaces, seashores, and life-sustaining
elements from the assaults of technology?

This is a book about man – about the devastating impact
of his numbers on the environment and the decisions and
actions he can take to attack the problem. The author is
an expert on conservation and planning. Land, air, water,
and wildlife are treated by him as both valuable resources
in very short supply and as precious living entities. He
contrasts present management of these resources with
man's future needs. British experience and examples from
all over the world illustrate the critical and practical
aspects of the problem. Past conservation programmes are
reviewed and evaluated, and the book offers a complete
set of proposals for regional, national, and international
action on environmental protection. The approach is
farsighted, informed, urgent.

a Pelican Original

Sleep

Ian Oswald

What is the nature of that mysterious condition, called sleep, in which we pass one third of our lives?

What part of the brain controls it? What happens when we give way to it? How much of it do we need? What takes place when we are deprived of it? Are there different kinds of sleep, and is it possible to learn while asleep?

And dreams . . . ? Do we dream at all hours? What is the explanation of dreams, and how quickly do we forget them? Do the blind see in their dreams?

And what, finally, are the newest findings about hypnosis and insomnia?

One of the most inexplicable rhythms of life is explained – as far as modern research can explain it – in this new Pelican by a psychologist who has specialized in the study of sleep.